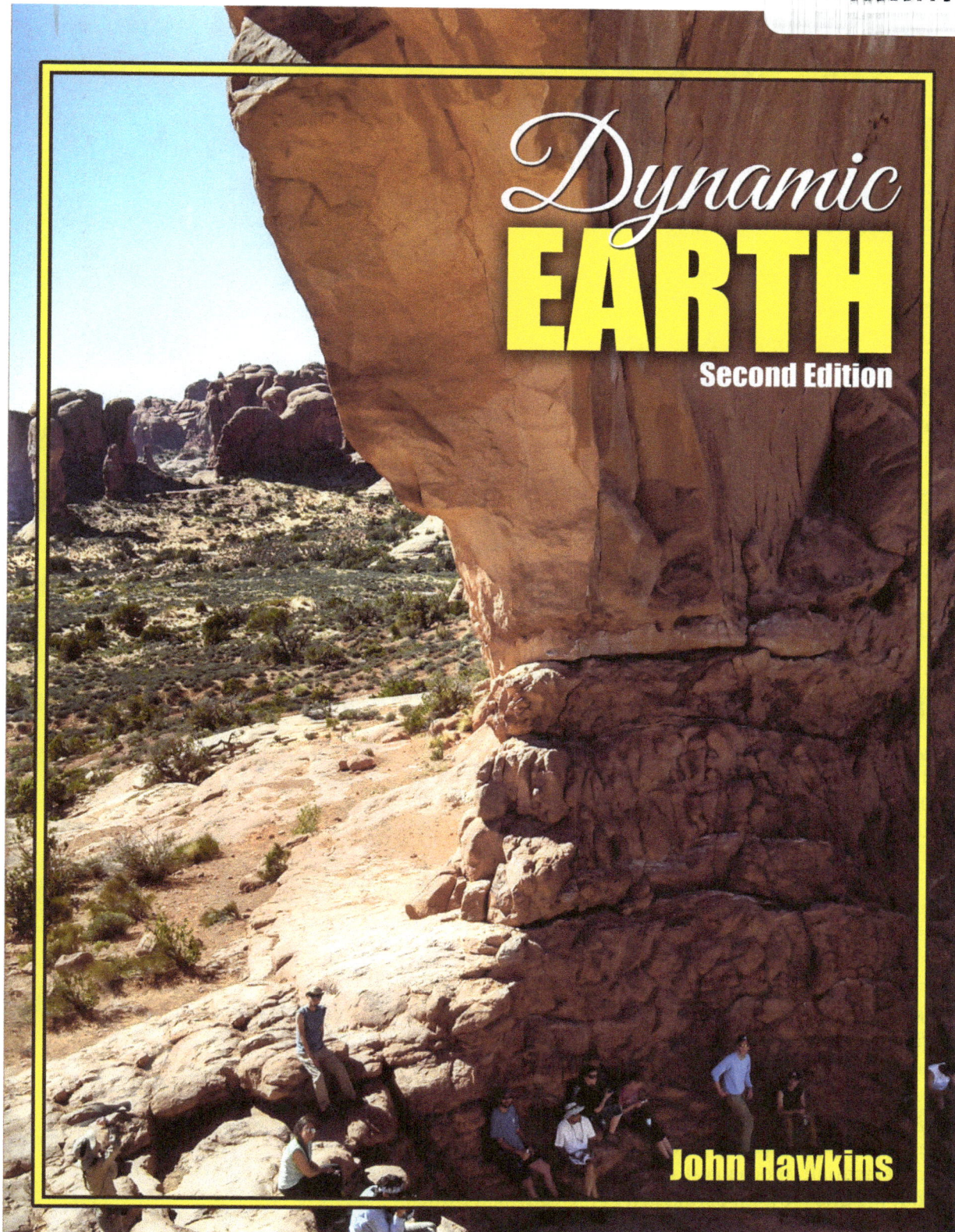

Dynamic EARTH

Second Edition

John Hawkins

Kendall Hunt
publishing company

Cover and interior photographs courtesy of John Hawkins.

Kendall Hunt
publishing company

www.kendallhunt.com
Send all inquiries to:
4050 Westmark Drive
Dubuque, IA 52004-1840

CONTENTS

CHAPTER 1

INTRODUCTION

WHAT IS GEOLOGY?

Now that you find yourself sitting in an introductory geology class, we need to start with the obvious.

What is geology?

The most commonly given answer to this question is the study of rocks. Okay, fair enough, geologists do spend a great amount of time studying and analyzing rocks. We study rocks in the field, we bring them back to our labs, we look at them under microscopes, and we may even shoot lasers at them in order to learn what secrets the rocks may contain. Sure, rocks are cool. However, geology is so much more than the study of just rocks.

Now back to the main question: What is geology? The word itself will actually give it away. In Greek, the prefix *geo* means Earth, and the suffix *ology* meaning a subject of study, leads us to the understanding that geology is actually the study of the Earth. There are many other things that make up our Earth other than just rocks. The career of a geologist will have many opportunities to explore the world around you. The box to the right provides a small list of subdisciplines that are available under the umbrella subject of geology.

A short list of careers in geology:

Volcanologist
Seismologist
Mineralogists
Petroleum geologists
Paleontologists
Hydrologists
Sedimentologists
Glaciologists
Environmental geologists

So that means you might find a geologist working on a volcano, on a glacier in Antarctica, at a gold mine, on an oil rig, testing the groundwater for your city, researching ancient fossils, or just about any other place you can imagine. The subject of geology will also utilize many aspects of biology, chemistry, and physics. If you enjoy learning science, world travel, and an adventurous lifestyle, maybe geology is the career path for you. If you have any questions about how to pursue geology as a career, please feel free to ask me or your TA during this semester.

As you probably noticed when you signed up for this course that there are two intro geology classes taught at Auburn University, and these are *Dynamic Earth* and *Earth and Life Through Time*. This is typically how the subject of geology is approached. It is divided into two broad areas that will cover the physical and historical aspects of the subject. You are currently taking *Dynamic Earth*, and this class will primarily be focusing on physical geology. **Physical geology** is the study of the processes that are actively shaping our world, and the materials that make our planet. When we say process, we are talking about things such as the rock cycle, weathering, erosion, or plate tectonics, but materials refer to the tangible parts of the Earth. We can sum that up simply by calling them rocks and minerals. When most people hear the terms rocks and minerals, they often think of something resembling figure 1.1. This is a collection of rocks and minerals that one would readily see at a gem/mineral show. These are generally exceptional examples and unfortunately are

Figure 1.1 **Rocks and minerals labeled and priced for sale at a mineral/gem show in Montgomery.**

not the types of rocks and minerals that a geologist would deal with on a regular basis. We will discuss rocks and minerals in detail in the upcoming chapters.

Historical geology is the study of how the Earth was made and how the Earth and life on it has changed or evolved during the long history of the planet. Geologists who specialize in past life-forms and their evolution are called paleontologists. Paleontologists are often found in some of the most remote places in the world, from the large islands off Italy to deep in Mongolia, looking for small clues to help answer the question as to how life has evolved on our planet. Figure 1.2 shows the west coast of the Antarctica Peninsula. The small inset photograph depicts small trace fossils that were created by small shallow sea creatures living there when the climate in Antarctica was much different from what it is today.

Earth and Life Through Time will focus on what these changes are and also the processes behind the changes. For example, to study the Appalachian Mountains would mean that a geologist would

Figure 1.2 Image taken from the west coast of the Antarctica peninsula facing north. Small inset shows trace fossils that were found at the photographs location.

have to break it down into the many events that acted over time to produce what we see today as the Appalachians. This would require a background in both physical and historical geology. A

union of these two broad areas is critical for one to become a successful geologist. The image in figure 1.3 will help to illustrate this point. In one image, we see the preserved remains of an ammonite, a shelled creature that lived in the oceans during the Mesozoic Era (251–65.5 Ma). The level of detailed preservation will allow paleontologists to learn a great deal about growth, life, and death of this organism.

Figure 1.3 Cut and polished ammonite fossils. This is an example of a commonly seen fossil sold for commercial purposes.

Upon closer inspection, you can see that many of the once void chambers are now filled with a myriad of colors. These wild colors within the ammonite are caused by different minerals.

After this creature died, fluids filled the once empty voids and allowed for minerals to precipitate and fill the regions. It would require an understanding of physical geology to be able to properly identify the minerals and the processes by which they formed. In this class, we will be discussing this particular mineral, quartz, in upcoming chapters. At that point, all the different colors will be explained.

The above image is a beautiful example of a fossilized ammonite. Most paleontologists would be ecstatic to have all their specimens with that level of preservation. Now before all of you change majors and become paleontologists, figures 1.4 and 1.5 illustrate what the majority of fossils do end up looking like. On the left, we have the backside of the preserved ammonite fossil from above. On the right, we also have an ammonite fossil with a poor quality of preservation. The degree and mode of preservation can make a big difference, as these two photographs show. Modes and methods of preservation will be discussed in the *Earth and Life Through Time* portion of geology.

Figure 1.4 The reverse side of the ammonite fossil in figure 1.3 showing excellent preservation detail.

Figure 1.5 Fossilized ammonite found in Montgomery, Alabama.

Who are geologists?

Anyone can become a geologist. Many jobs in geology are based on fieldwork, but there are also many that are laboratory based. Many people have the misconception that most geologists are old bearded men who swing hammers at rocks, and in some cases that is true. In recent years, the field of geology has enjoyed an expansion of young professionals as well as colligates. When attending a national conference that covers most topics in geology, it appears that the younger generation of geologists is near 50/50 men and women. It does appear that the secret is out. More and more students are learning about the adventurous and lucrative careers that are readily available to geology graduates.

The person in figure 1.6 is one of our Auburn University graduate students in the field conducting field observations. You can clearly see some of the tools that are common for a geologist to have while conducting a fieldwork. He has his waterproof notebook for recording his observations, and you can see his hammer down by his feet. This is for breaking off smaller rocks so that the samples can be brought to the laboratory for research, and sometimes it is also used to break away material so that the geologist can see a fresh surface on the rock.

Many of you have probably seen geologists out conducting fieldwork or in a class where they are learning the necessary skills to be a field geologist. Often they are along roadsides inspecting and measuring the orientations of rocks that were recently exposed during road construction. If you see rocks, then more likely than not, at some point a geologist has been there. You have probably passed by many interesting rock outcrops in your travels and never noticed them. Once you finish the course, you will

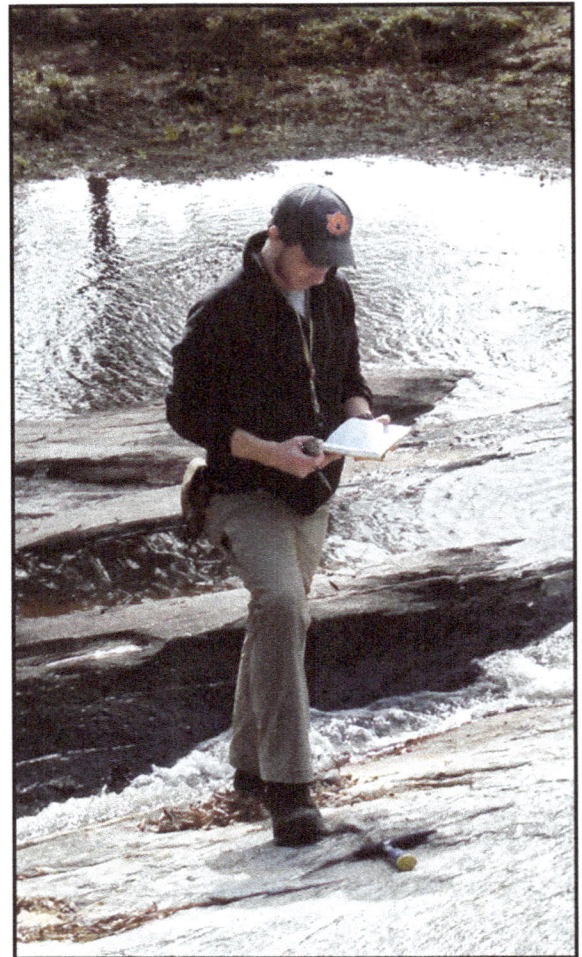

Figure 1.6 A graduate geology student in the field.

probably see many rocks that you have previously taken for *granite*. Okay, there are not a lot of great jokes in geology, so we do the best we can. The below photographs show what some typical fieldwork might look like as well as some of our past students out learning geology fundamentals.

Geologists in the field

Figure 1.7 Inspection of a recently exposed rock face near Dadeville, Alabama.

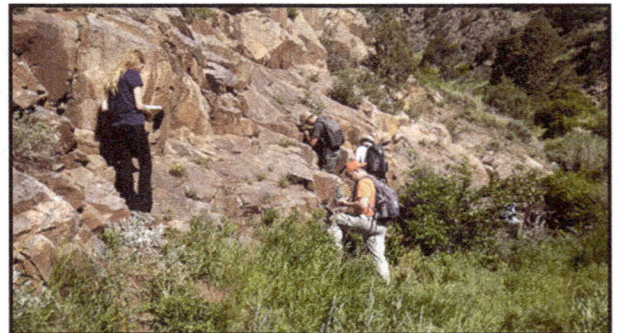

Figure 1.8 Students learning basic field techniques on exposures found in Denver, Colorado.

Earth's System

At this point you should be realizing that the field of geology is much more rich and diverse than you previously thought. Geologists study the Earth, all of it, and in order for us to be successful, we need to view the Earth as a system. A **system** is a collection of interacting, or mutually supporting, parts that form a complex whole. The best example of a functioning system is your body. All the organs and tissues that work in tandem to allow you to sit there and read this page

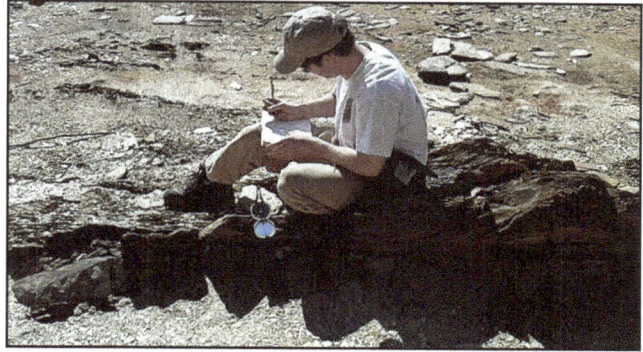

Figure 1.9 **A graduate student sitting on an outcrop near Dadeville, Alabama, recording field observations for his thesis.**

are an example of individual parts making up a whole. If any of these parts stopped working, the entire system would be affected.

The Earth operates in a manner very similar to our body. The Earth has spheres, similar to organs in our body. The Earth has four interconnected spheres that work together to create the *dynamic* planet that you live on. These four spheres include (1) the liquid and solid water component, the *hydrosphere*; (2) the gases that surround our planet, the *atmosphere*; (3) the solid rocks and materials that make up the Earth, the *lithosphere*; (4) and lastly, all life-forms on the planet make up the *biosphere*.

All four of these spheres are constantly interacting and the level at which they are connected is incomprehensible. Figure 1.10 will help demonstrate this connection.

Figure 1.10 **Late evening on an island off the coast of Antarctica.**

In figure 1.10, three of the four spheres are clearly visible. You can see the rocks of the lithosphere, the snow and ice of the hydrosphere, and the clouds within the atmosphere. In this location it is easy to see how the lithosphere interacts with the hydrosphere, where land meets the water. Most shorelines are excellent examples of this, but in this image, the connection goes further. The mountains in the background are covered with snow and ice. They are constantly being shaped by the hydrosphere. Those snow and ice are generated by the atmosphere and its interaction with the hydrosphere. The lithosphere interacts with the hydrosphere and the atmosphere by blocking and channeling the flow of the atmospheric gases. This is what brings rain to some regions and creates deserts in others.

Hydrosphere

The hydrosphere includes all the liquid and solid water on the planet (figure 1.11). This important liquid is in a constant state of change due to evaporation, condensation, precipitation, and back again. All of you have learned and studied the hydrological cycle at some point during your education. This is a major component of the dynamic Earth system. The hydrosphere covers 71 percent of our planet's surface. Of this 71 percent, approximately 97.5 percent is saline (salt) water leaving only 2.5 percent as freshwater. This 2.5 percent freshwater includes groundwater, streams, lakes, and glaciers.

Figure 1.11 **Natural ice arch within icebergs off the coast of Antarctica.**

Atmosphere

The atmosphere is a thin layer of gases that surround and envelope our planet (figure 1.12). Not only does the atmosphere provide elements necessary for life to produce energy, it also protects the planet from harmful radiation that spills out from the sun and floods into our solar system. Our atmosphere is 78 percent nitrogen, 21 percent oxygen, and 1 percent others (water vapor, carbon dioxide, carbon monoxide, argon, neon, sulfur dioxide, methane, and ozone).

When compared to the diameter of the
Earth, about 4,000 miles, the atmosphere
that interacts with the other parts of our
system is rather thin. The atmosphere is
divided into five zones, and they are (1) the
troposphere, this includes the near surface
gases and the majority of our weather;
(2) the stratosphere; (3) the mesosphere;
(4) the thermosphere; (5) and the exo-
sphere, which is considered to extend as
far as 6,000 miles above the surface. Even

Figure 1.12 **Dramatic sunset on the Isle of Skye, Scotland.**

though there are gases associated with our atmosphere that extend to the 6,000 mile mark, the
majority of all the atmospheric gases are found within 30 miles of the surface.

The column of atmospheric gases exerts a pressure on the surface. One common way to express this pressure is with the term **atm** (atmospheres); 1 atmosphere is the pressure exerted
by this column of gases as felt at sea level, and it is near 14.7 lbs/in.^{2} On the Auburn University
campus, we are at an elevation that is between 500 and 600 ft above sea level. That means that
most of us are used to an atmospheric pressure of around 1. The higher in elevation that we
travel, the lower that number will become. For example, on the top of Mt. Everest (29,029) the
atm would be 0.3. Only 30 percent of the gases that are available at sea level can be found at
that elevation.

Biosphere

The biosphere includes all life on
the planet (figure 1.13). The majority of life on Earth can be found
from 3 m below the ground up to
30 m above the surface, and the
top approximately 200 m of the
oceans. The majority of life is confined to a narrow band around the
planet. There are many extremes to

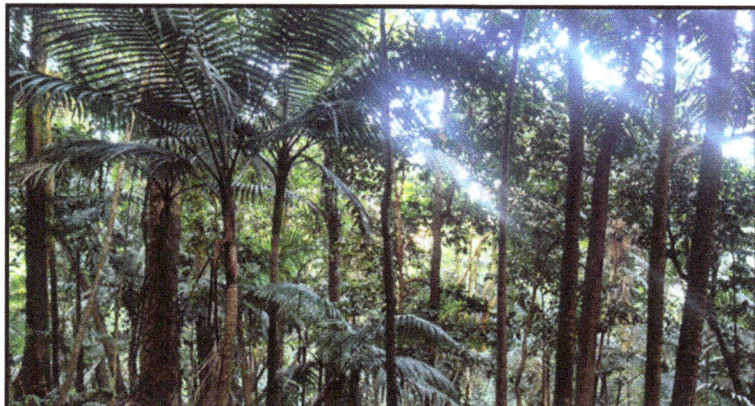

Figure 1.13 **Biodiversity of the El Yunque National Forest in Puerto Rico.**

Figure 1.14 An estimated one million mating pairs of kind penguins on the island of South Georgia in the southern Atlantic.

where life can exist (figure 1.14). It has been found existing as deep as 2.5 miles below the surface and as high as several miles up into the atmosphere. Even with these extreme examples, life is still limited to a narrow region. This is a geology course and we will not be studying the biosphere in great detail. We will be learning about how the biosphere interacts and affects the other spheres in the Earth system. The interaction of life, from humans to microbes, is very complex and plays a large role in the evolution of the other three spheres.

Lithosphere

The lithosphere, or geosphere, extends from the surface down to the center of the planet, and includes all the solid material and molten rock within the Earth (figure 1.15). The Earth is subdivided into four different layers: the inner core; the outer core; the mantle; and the crust. This internal differentiation is critical to the operation of the Earth system. The specifics and mechanisms for this differentiation will be discussed in class.

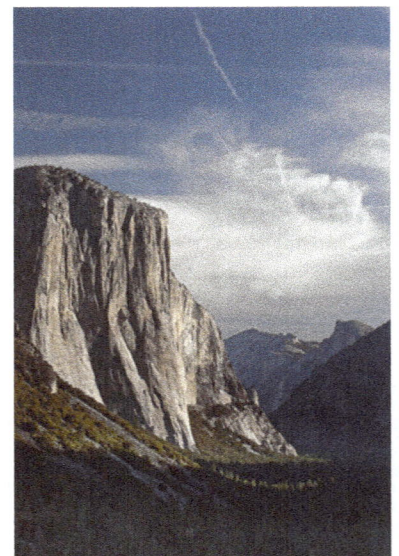

Figure 1.15 El Capitan with Half Dome in the distance.

The rocks and processes acting on and within the thin (10's of km), solid, brittle, rigid, and cool crust will make up the majority of this course. The crust is also broken into many different plates. We will cover the complex interactions of these plates (plate tectonics) in later chapters.

The crust can be divided into two different broad categories. One is oceanic crust, this is crustal material that is generally found at the bottom of our oceans, and the other is continental crust. As the name implies, this is the crustal material that our continents are made of. Generally, continental crust has a lower density than oceanic crust and this allows it to essentially float on top of the oceanic crust. Continental crust is rich in **felsic** (Si and Al) minerals while oceanic crust is more **mafic** (Fe, Mg, Ca, and Na) in composition. This compositional difference will be discussed in the upcoming chapters.

These compositional differences will largely affect **viscosity** and the types of minerals that we will find in each type of crust. Viscosity refers to the resistance of flow by a material. If a material has a high viscosity then you can think of it as being thick and slow moving. So logically, if a material has low viscosity then it will be thin and runny. The viscosity is controlled by the amount of Si present in the material. Felsic material is high in Si and therefore it will be more viscous than mafic material.

Concept Review

In order for geology to function as a science, geologists must approach each new problem or unknown with the scientific method. This method of problem solving has been covered in every science class you have ever taken. In class, we will talk about a few of the finer points and how they directly apply to the study of geology. Instead of going through each step of the scientific method, I am going to list the process here for you.

Instead of lecturing in detail about the scientific method, we are going to work through as an

Scientific Method

A. **Define the problem or ask a question**

B. **Collect and evaluate data**

C. **Hypothesize (make an educated guess to explain your observations)**

D. **Test of time –Does your hypothesis work over time?**

assignment. I am going to give you a specific question. This will be your point A (starting point). From there I would like you to list any data or observations that you have made or that you are aware of about the question as point B. For point C, I would like you to make an educated guess at the answer. Leave point D blank and we will talk about it in class. Your question is *"Does the moon rotate on its axis?"*

Once a scientist has worked through the steps of the scientific method, and they have produced a valid time testing hypothesis, it can then be called a theory. A theory is a scientific idea supported by an abundance of evidence that has passed many tests and failed none. There are several theories that we will talk about in the course. Plate tectonic and evolution are two theories that are very prevalent in geology. Since both of those are valid theories, it means they have never failed a test, and they also allow for science to make predictions about the future.

When a theory becomes accepted as a natural law, it becomes a principle of science that must be absolutely true. If this principle were violated, the universe/physical world as we know it would not exist. The Law of Gravity, Law of Original Horizontality, and Law of Superposition are examples that we will be talking about in upcoming chapters.

Questions

1. What is the average salary of a geologist with a bachelor's degree?

2. What is the average salary of a geologist with a master's degree?

3. What are the two broad categories of geology?

4. Why would someone need an understanding of both categories of geology to be a successful geologist?

5. What are the four spheres in the Earth system?

6. Describe a situation/place where all four spheres are interacting.

7. What is viscosity?

8. What is geology?

CHAPTER 2

ROCKS AND MINERALS

MINERALS/BOWEN'S REACTION SERIES

Чароит
Якутия, р. Чара
Федореев В. Н. 2012 г.
орбр. 342   кол. 576

What is a rock?

Now that we know geologists actually deal with more than rocks, we actually need to start this course talking about that very thing, rocks. If you ask someone to go outside and find a rock, it would be an easy task. Most of us encounter rocks on a daily basis. They can be found around most homes where they are used in various landscaping, and most people are aware that they are in our cemeteries. Rocks can outcrop in rivers, like we see on the opening page of this chapter, or they can seem very dynamic, like the rock on the top of this page, with swirling colors of purple and green. If you then ask this same person to define what it means to be a rock, you might get a different reaction.

In addition to landscaping, we often use many various types of rocks in our homes (figures 2.1 and 2.2). The most common locations are flooring and stone countertops. Let's take a look at a typical granite countertop and see if we can formulate an answer to the question, **what is a rock?**

Figure 2.1 Grave marker for the famous Antarctic explore, Shackleton.

14

Figure 2.3 is a photograph of a typical granite countertop like you would find in many homes. Yes, by definition granite is a rock. So let's take a closer look. In the above photograph, do you see different colors? There are shades of gray, white, black, and some red visible in that example of granite. All the different colors you see represent a different mineral, and those minerals together make up the above rock. In the above example, there are easily four to five different minerals present. **A rock can be loosely defined as a large mass of a single mineral or an aggregate of different minerals**.

Okay, rocks are a collection of minerals. That is easy enough; in figures 2.4 and 2.5 we have two different examples of rock types. In figure 2.4 we can see some large examples of a dark reflective

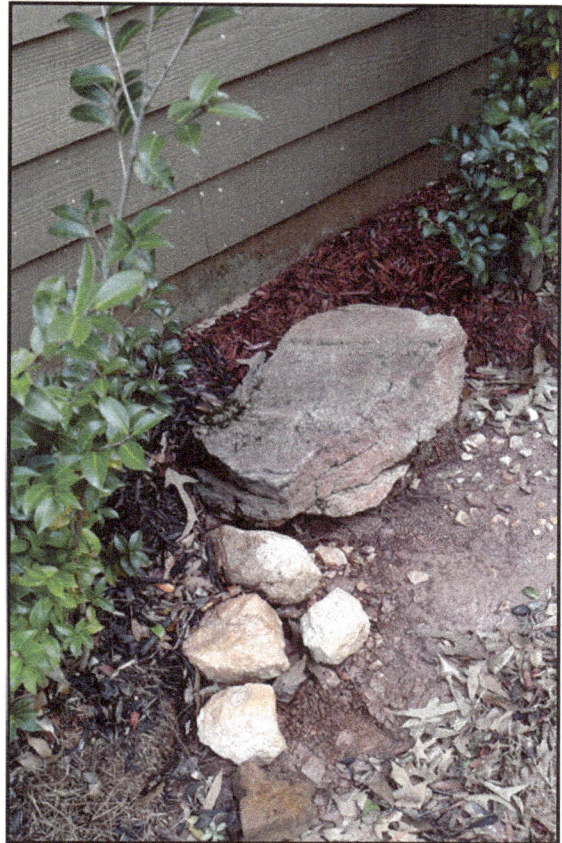

Figure 2.2 **Typical local rocks used as landscaping.**

Figure 2.3 **Granite countertop.**

Figure 2.4 **Rock with sphalerite.**

Figure 2.5 **Quartz.**

mineral called sphalerite. That is an example of one mineral being very prominent in a rock sample with a collection of different smaller minerals filling in the matrix behind the larger sphalerite crystals. In figure 2.5 we see the mineral quartz. Everything in that image is quartz. So what we have is an example of a rock made totally from the mineral quartz. Now we need to define what is necessary for something to be called a mineral.

What is a mineral?

Determining if a substance is a mineral is based on five criteria. You can ask these five questions about any substance and as long as all the answers are yes, it's a mineral.

1. Is it a solid?

Now this is an easy one. Determining if a substance is in a solid state is usually very apparent. If it is in a solid state then it passes the first requirement. I do want you to think about the natural phase shifts that many materials go through. Just because a material is in the liquid phase at STP (standard temperature and pressure) does not mean that it will not become a solid once the temperature or pressure has been changed. So with that being said, you can have minerals that exist in some pressure and temperature ranges that do not exist as minerals in others.

2. Is it inorganic?

This means that to be considered a mineral, the substance must not be made from organic compounds. The definition of an organic compound can sometimes be a tricky one. It is generally

Figure 2.6 Bay off the coast of Antarctica that is collecting iceberg remnants.

assumed that if a molecule contains C then it is organic. This is not the case in geology. Many of our minerals, such as calcite, aragonite, siderite, and dolomite, are considered inorganic, although they contain C in their structure. In this course, we revise the definition to require a C–H or C–C bond to be considered organic.

3. Is it naturally occurring?

For something to be considered naturally occurring, the substance must be formed by natural geologic processes. This statement is used to differentiate materials that are derived synthetically. Materials that are created by human interventions are not considered minerals.

4. Does it have a definite chemical composition?

Minerals, like everything else, are made of atoms. These atoms form chemical compounds and we identify specific minerals by their chemical formula. For example, the mineral olivine has a chemical formula of $(Mg,Fe)_2SiO_4$. I am sure you remember from earlier chemistry classes that when we write the chemical expression with parenthesis, it means that either of the elements can be used in that chemical formula. Olivine can be Mg_2SiO_4, Fe_2SiO_4, or $FeMgSiO_2$ with varying percentages of both Fe and Mg. When we say that the mineral has to have a definite chemical composition, we mean that olivine will have that same formula today, tomorrow, and hereafter. Olivine will not all of a sudden be defined by K_2SiO_4.

5. Does it have a crystalline structure?

This means that the atoms making up the minerals are arranged in a set orderly fixed pattern. This forms a repeating pattern that is expressed as crystals. Each time this mineral grows, the atoms will arrange themselves into this fixed pattern as long as there is room and time to do so. In figure 2.7, you see a crystal lattice structure formed from the bonding of sodium (Na) and chlorine (Cl). This is represented by the purple and gold in the ball-and-stick model. You can clearly see the cubic structure that is formed by the bonding of NaCl. When these two atoms bond and form this crystal lattice, we call the formed mineral halite. Figure 2.8 is an image of halite. You can see the expression of the cubic lattice in the shape of the mineral.

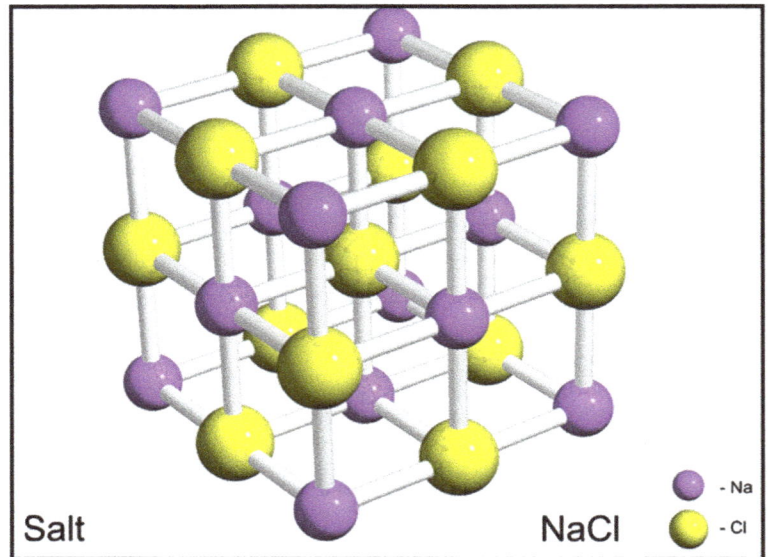

Figure 2.7 Crystal lattice structure.

How do minerals form?

When molten material cools or crystallizes it forms rocks and minerals. Now not all minerals are produced from cooling magma, but understanding how these minerals form is a good place to start. We will move on to some other types later in this chapter.

Figure 2.8 Halite.

Now we need to clarify the difference between the terms magma and lava. Both are forms of molten rock, but magma is molten rock that is found beneath the Earth's surface. Lava is a term we use when magma reaches the surface. If you see molten rock material, you are looking at lava. It is within the lava and magma that rocks/minerals can form. Later in this course we will discuss the difference in the rocks formed by magma/lava, but for now we will just focus on the process.

So imagine a large mass of molten rock or magma. Within this mass there are many elements that can combine to form minerals. There are eight elements that we focus on as the building block for the rock-forming minerals. The formation process that we are about to discuss applies to minerals; we call them rock-forming minerals. As their name implies, these are the main minerals that form rocks.

Of these eight elements, the first two we will talk about are among the most abundant elements within the Earth's crust, Silicon (Si) and Oxygen (O), and they are among the first elements to bond in magma/lava. In order to understand how these two elements will combine we need to talk about valence electrons. Think about valence electrons as the electrons in the outer shell of the atom that are the ones that interact with other atoms during bonding. At the risk of oversimplifying the process, we will consider an atom to be content when it has a full outer shell. In the case of O and Si, this can be achieved in one of two ways. Si, in its neutral state, has four electrons in its outer shell. It can either steal four others to complete its outer shell or it can give those four electrons away. Si, in this case, gives them away. So Si is considered to have a +4 valence number indicating that it has four to give away. Oxygen has six electrons in its outer shell and it is much easier to gain two more and complete the outer shell instead of giving away all six. So O is considered to have a −2 valence number indicating it wants two electrons. When an atom either gives away or gains an electron, we call that atom an ion. Ions that have a positive valence number, in the case of Si, we call those cations, and ions that have a negative valence number, in the case of oxygen, we call those anions.

When Si and O start to bond, they form a structure called a tetrahedron. It is made of four oxygen atoms and one silicon atom. Figure 2.9 shows the structure of the tetrahedron with Si in black and O in light gray.

Now we need to do some quick math. We need to figure out if the SiO tetrahedral has an overall charge. Four oxygen atoms with −2 each have a negative charge of −8. We also have one silicon atom with a charge of +4. So the overall charge on the tetrahedral is −4 (−8 + 4 = −4).

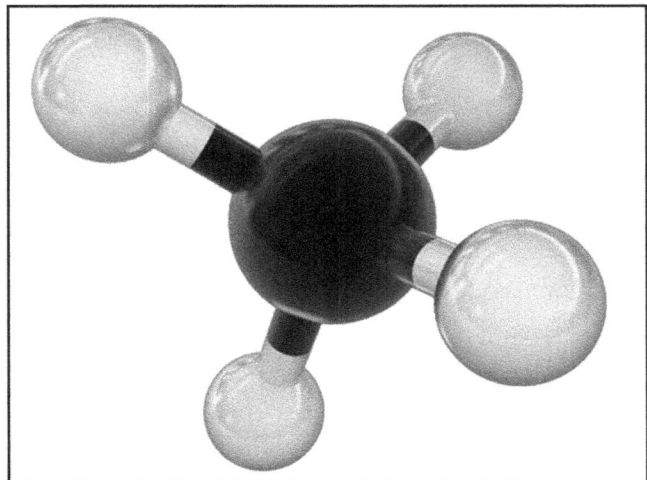

Figure 2.9 **Tetrahedron.**

Just when we need it, 1980's song lyrics come to the rescue with wise words about opposites attracting. With the tetrahedron having a -4 charge, it will attract other cations to bond and neutralize the negative charge. When these other elements bond to neutralize this negative charge, a mineral is formed.

In the melt we mentioned eight elements earlier, and we have already talked about silicon and oxygen. This leaves six atoms that are common in the melt that we can have bond with the tetrahedral to neutralize the -4 charge. These other six atoms are cations and they all have varying valence numbers.

For example, two iron atoms can bond with the tetrahedral and their $+4$ charge would balance out the -4 of the tetrahedral. This could also be done with two magnesium atoms or one iron and one magnesium atom. If you remember back to earlier in this chapter, you will recall that we gave the chemical formula for a mineral named olivine (Mg_2SiO_4).

2 Magnesium atoms + 1 Silica tetrahedron

$$Mg_2 \quad + \quad SiO_4$$
$$+4 \quad + \quad -4$$

Olivine

This process of bonding cations with the negatively charged tetrahedral is the process in which we make minerals. The different combinations of cations that can bond to neutralize the negative charge will create many different mineral formulas with many different crystal lattice structures. This allows for other rock-forming minerals to be formed other than just olivine.

How do we form minerals beyond olivine?

Now that we are forming the chemical bonds to make olivine, let's think about what is happening within the melt. Imagine many tetrahedral olivine structures floating in the liquid melt material. Figure 2.10 shows many isolated olivine tetrahedrals. The key word here is isolated. Each isolated tetrahedral can be thought of as a single

Figure 2.10 **Many isolated olivine tetrahedrals.**

olivine mineral moving rapidly around in this melt. The rate of this motion is related to the temperature of the melt. The higher the temperature, the faster the molecules are moving. The tetrahedrals are moving rapidly and in order for them to be able to bond to each other the temperature needs to drop. Magma cooling will result in more bonding, and more bonding will result in different minerals. Figure 2.11 will illustrate this process. The key idea to remember is that if we are bonding tetrahedrals, then we are changing the structure.

Random assortment of isolated neutral tetrahedral structures.

As the temperature drops, bonding formssingle chain structures.
*we now have a different crystal structure

As the temperature continues to drop, we form double chain structures.
*we now have a different crystal structure

As the temperature continues to drop, the double chains from sheet-like structures.
*we now have a different crystal structure

DECREASING TEMPERATURE

Figure 2.11 Process of magma cooling.

That means that the fifth criterion for forming a mineral is now very important. If we change the structure, we have made a new mineral.

Each of the above crystalline structures relates to a different mineral. We see four different structures, so we will have four different minerals. The isolated tetrahedral we know is olivine. When it bonds and forms a single chain, the mineral formed is called pyroxene. Pyroxene will have a different set of properties from those expressed in olivine. We will talk about mineral properties in detail later in this chapter. You can now think of the basic mineral property of color for this example. Olivine with its greenish color will differ from the darker color of pyroxene due to its different properties. The important idea here is that single tetrahedral olivine bond to form a new mineral. Once this bonding has occurred, it is no longer olivine, but it is now pyroxene.

As the magma continues to cool, the single chains of pyroxene will bond to form a double chain. This forms the mineral amphibole. With the cooling continuing, these double chains will bond and form a sheetlike lattice, and we call that mineral biotite. So at each step we make new minerals by making changes to the crystal lattice. At high temperatures we are forming olivine

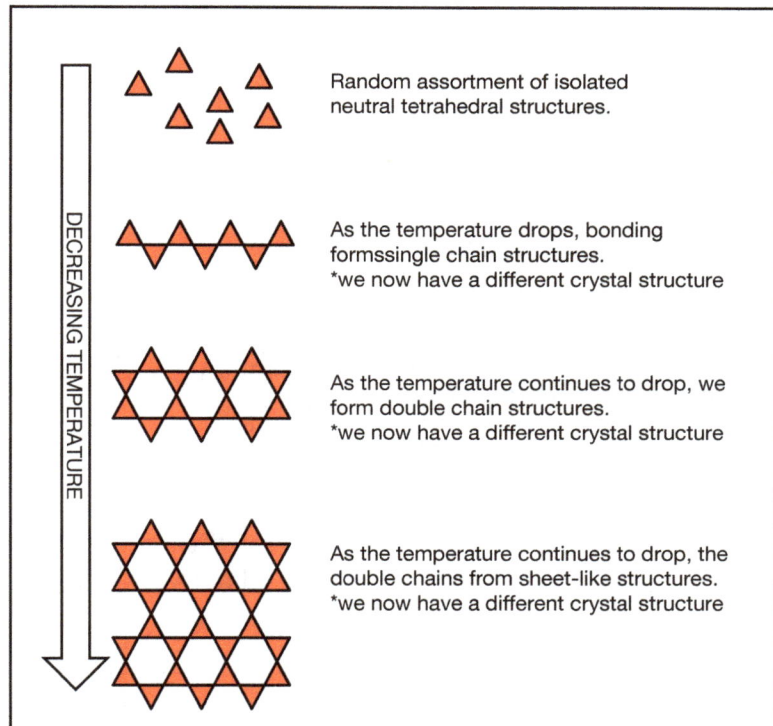

and as the temperature drops we will eventually be able to make biotite. This process is temperature dependent (figure 2.12).

Earlier we mentioned six cations that were abundant in molten material that are used to make these minerals (Mg, Fe, Ca, Na, Al, and K). The above minerals use Mg and Fe, and we talked about how they are used in the formation of olivine. When minerals use Fe or Mg, we call these minerals mafic. The word mafic means darker colored minerals that contain Fe and Mg.

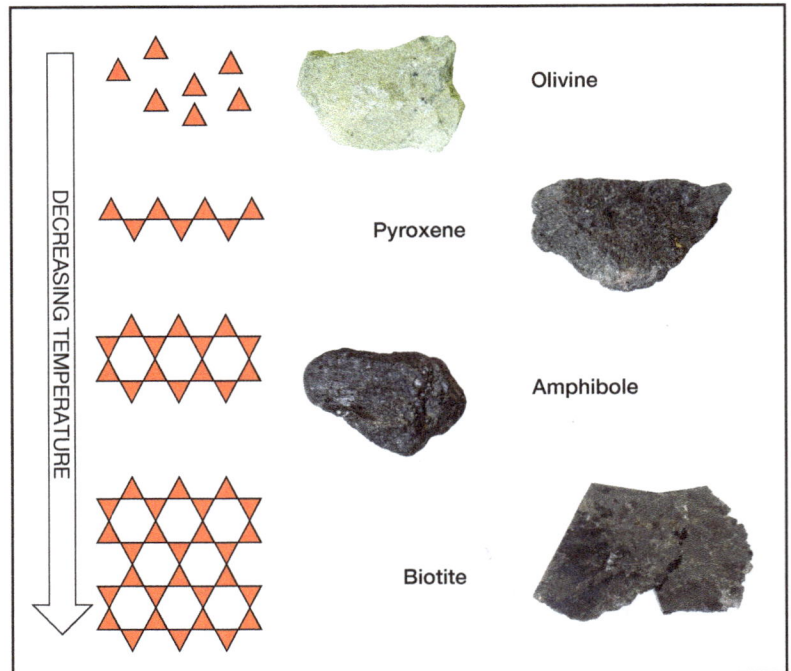

Figure 2.12 **The process of mineral formation is temperature dependent.**

This means the four above minerals would be considered mafic minerals. So what is happening to the abundances of our six elements if these higher temperature minerals are using Fe and Mg? That would mean that the concentrations of Ca, Na, Al, and K would be higher relative to the mafic (Fe and Mg) elements as the magma continues to cool. We do have a mineral that we have yet to talk about that will be using up the Ca. This mineral is called plagioclase, and you will have plagioclase in lab where you will talk about its specific mineral properties. Plagioclase also likes to form at high temperatures. It forms a little more complicated crystal lattice. Whereas olivine is based on a single Si and O tetrahedron, plagioclase is based on two combined tetrahedrons. So let's take a look at the root lattice of plagioclase and start with Si_4O_8. At this point, we need to briefly talk about elemental substitution.

In the crystal lattice of all minerals, substitutions can be made. This means that often different atoms can step in and fill the location that would normally be occupied by one of our six cations. This process of substitution will ultimately allow for use to make the abundance of minerals we find on the Earth. Substitution is only possible when atoms are of similar size and charge. For a real world example, we are going to use a high school classroom. The 4 walls, the 28 students, and the teacher make up the crystal lattice of Mrs. Jenkins 10th grade biology class.

Now imagine for a second that Mrs. Jenkins becomes ill. She is no longer available to fill her spot in the classroom/crystal lattice, and that leaves a hole or gap. You probably remember from high school that a gap like that is not going to be tolerated and a substitute is found to take her place. This person will be of similar size and perform a similar function, and the classroom can move on throughout its day.

Some substitutions have little effect on the overall lattice. If one of the 28 students was substituted for another, there would be very little disruption and little change to the overall lattice. This type of substitution does not change the mineral that is being formed. Other types of substitutions can be a little more drastic. In the case of substituting Mrs. Jenkins in her own biology class, once she is replaced it is no longer Mrs. Jenkins biology class and we have, in other words, formed a new mineral. So some types of substitutions do not change the mineral while others will. We will talk more about this concept in class.

In plagioclase, we started with Si_4O_8 (figure 2.13). We are going to do some substitutions that are similar to changing students in their desk. We are going to substitute two Al atoms for two silicon atoms. They are roughly the same size but their charges are different, Al (+3) with Si (+4). After the substitution, we have the start of a crystal lattice that is $Al_2Si_2O_4$. When we do the math on the valence electrons, we find that this structure is a −2. We need to find an element out of

Figure 2.13 Plagioclase.

our six that can bond and neutralize this charge. The Fe and Mg are busy with the olivine, and their size allows for them to be very happy over there. So our other +2 option will be calcium. Calcium is much larger than Fe and Mg, but it fits in the plagioclase structure very well. So we add one Ca and then we have the neutral structure of plagioclase—$Ca(Al_2Si_2)O_8$.

As the melt cools we end up depleting our supply of Ca. Now we need to find another element and do some other needed substitutions. We alter the $Al_2Si_2O_8$ to $AlSi_3O_8$ and that substitution changes the overall charge of the Si and O lattice to −1. The −1 charge is easily neutralized by adding a sodium atom (Na). This type of substitution, changing from Ca to Na, is equivalent to changing out students within our classroom. This does not change the mineral name. The reason is when we use Ca or Na, it does not alter the mineral properties enough to classify it as a different mineral. They are both called plagioclase, and we refer to

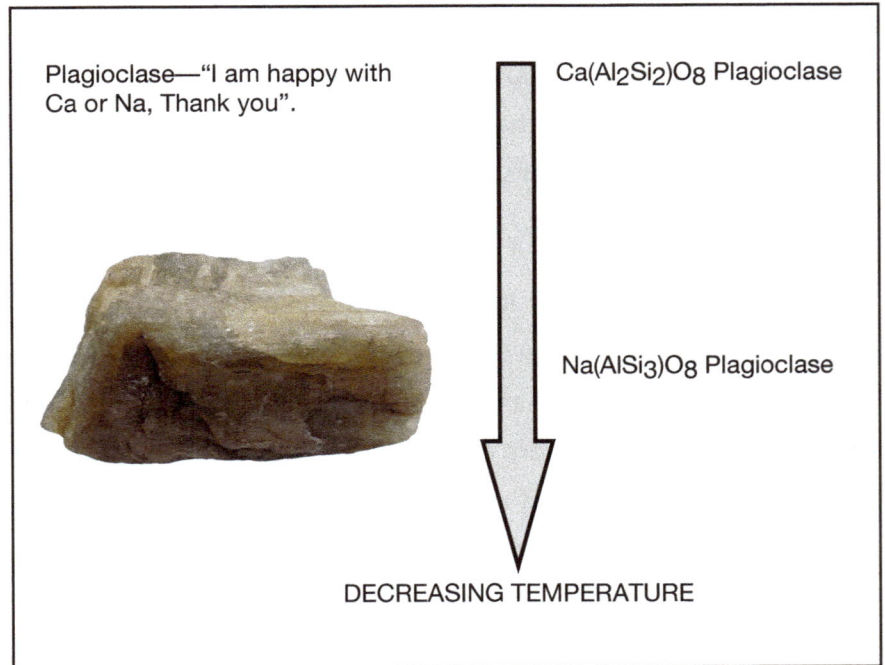

Plagioclase—"I am happy with Ca or Na, Thank you".

$Ca(Al_2Si_2)O_8$ Plagioclase

$Na(AlSi_3)O_8$ Plagioclase

DECREASING TEMPERATURE

Figure 2.14 Plagioclase can be Ca-rich or Na-rich.

them as Ca-rich plagioclase and Na-rich plagioclase. The Ca-rich variety forms at higher temperatures and as the melt cools, we transition to the Na-rich plagioclase (figures 2.14 and 2.15).

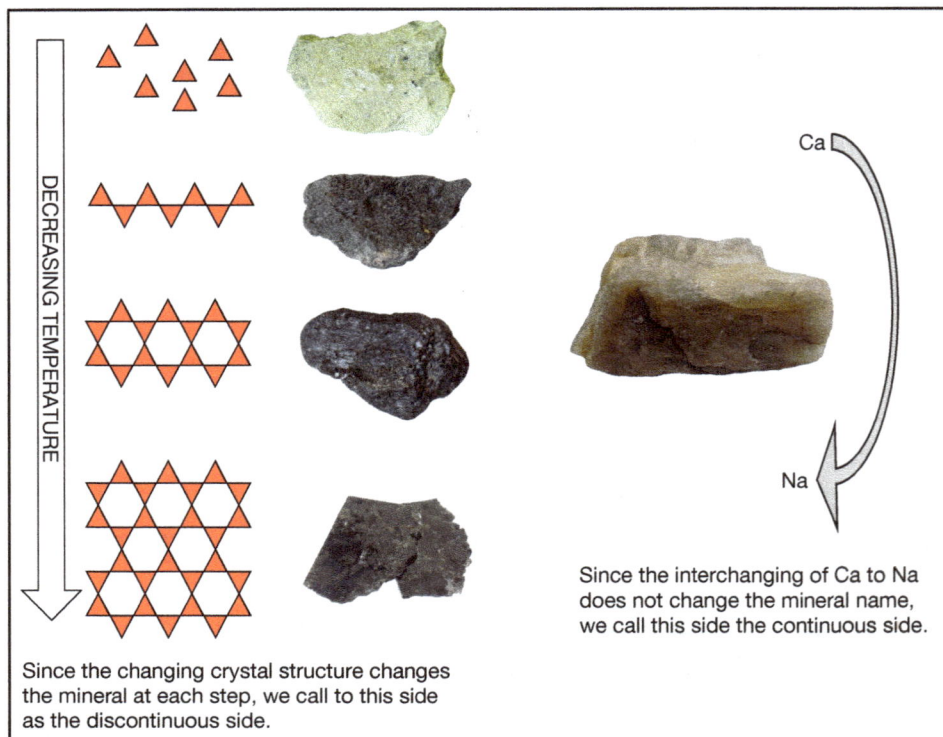

DECREASING TEMPERATURE

Ca

Na

Since the interchanging of Ca to Na does not change the mineral name, we call this side the continuous side.

Since the changing crystal structure changes the mineral at each step, we call to this side as the discontinuous side.

Figure 2.15 Discontinuous versus continuous sides.

Please keep in mind that the formation of plagioclase is happening at the same time as the other four minerals that were talked about earlier. So there are multiple minerals being formed within the melt.

So now we find ourselves forming five minerals as the melt is cooling (olivine, pyroxene, amphibole, biotite, and plagioclase) (figure 2.16). We have been using Fe, Mg, Ca, Si, O, and some Al. We are staring to run low on resources to make minerals, but we find that we still have high concentrations of K, Si, O, and Al. Si and O are the most abundant elements, and these guys are available to the very last, but we have used up all the Fe, Mg, Ca, and Na. So what minerals can we make using what we have left with (Al, K, Si, O)?

Figure 2.16 **Various minerals.**

We can continue to form a mineral that is very similar to Na-rich plagioclase. Recall its chemical formula—$Na(AlSi_3)O_8$. Since we are out of Na, we are left with $(AlSi3)O_8$ and that as we discussed earlier presents a -1 charge. The only option we have left to cancel out this -1 charge is to use potassium. This becomes $KAlSi_3O_8$. This is Potassium Feldspar or K-spar and it is very similar to the Na-rich plagioclase that we were making earlier; however, this particular substitution of K for Na does change the mineral properties a good deal. This is the type of substitution that we talked about earlier where we are changing Mrs. Jenkins for another teacher. The overall structure will be different due to the size difference from K to Na. So that is why it is a different mineral.

This mineral growth will start to deplete the potassium. Another mineral starts to form about the same time that will also be using potassium and aluminum. This mineral is called muscovite, we are not going to worry about its chemical formula but just know that it is forming around the time we are making K-spar. These two minerals, K-spar and muscovite, will consume all the Al and K. This leaves just Si and O in the melt. These two will form a modified shared tetrahedral structure SiO_2 (neutral) that forms the mineral quartz. So the last three minerals to form in this series are K-spar, muscovite, and quartz.

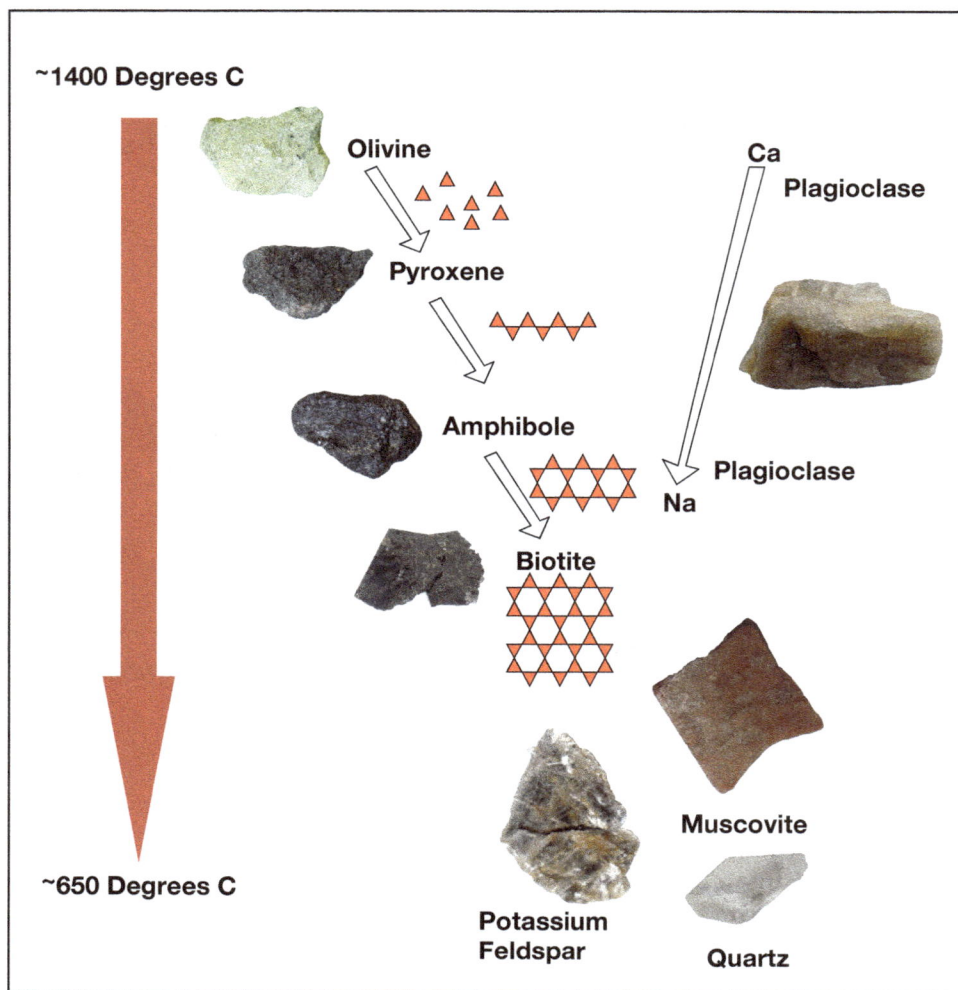

Figure 2.17 Bowens Reaction Series.

This sequence, starting with olivine and plagioclase and ending with quartz, is called **Bowens Reaction Series** (figure 2.17). This is a critical concept to any young geologist. You need to go ahead and memorize this sequence now because it will be utilized in many topics throughout the semester. This process produces eight minerals from a liquid melt. These eight minerals are considered to be the main rock-forming minerals, and since all of these minerals contain Si and O, they can collectively be referred to as **silicates**. To be a silicate mineral, the structure/chemical formula must contain Si and O. In addition to this rock-forming silicate group, we will be discussing six other mineral groups in class. These groups do not contain Si and O and are called **non-silicate groups**. These groups are generally not thought of as rock-forming minerals. Minerals in these six groups generally occur as minor **accessory minerals** within rocks.

Mineral Properties

Now that we understand how our main rock-forming minerals and accessory minerals are created, we need to talk about how geologists tell these minerals apart. Each mineral will have its own distinctive set of unique mineral properties. These unique properties can be used to identify that specific mineral from all other minerals. No two minerals will have exactly the same mineral properties, and you can think of these properties forming something like a mineral fingerprint. In class, we will talk about several of the mineral properties that will be important for the class. In lab, you will also work to identify silicate and non-silicate minerals using these distinctive properties.

- Luster
- Hardness
- Cleavage/Habit
- Fracture
- Color
- Streak
- Specific gravity
- Magnetism
- Taste
- Odor
- Feel
- ACID
- Other

Color and Density

In addition to the silicate and non-silicate mineral distinction, there are two broad categories based on the Fe and Mg content within the silicate mineral group. These are **ferromagnesian** and **nonferromagnesian** silicates, which simply means iron and magnesium bearing or not. The presence or absence of Fe and Mg strongly affects the external appearance (color) and density of the minerals.

The distinction between ferromagnesian and nonferromagnesian will become increasingly important as we move into the upcoming chapters. Look back over Bowens Reaction Series and make sure you can associate each mineral with its correct category.

Ferromagnesian silicates—dark color

- **Olivine**

- **Pyroxenes**

- **Amphiboles**

- **Biotite**

Nonferromagnesian silicates—light color

- **Muscovite**

- **Plagioclase and K-spar**

- **Quartz**

Questions

1. What is the definition of a rock?

2. What is a mineral?

3. What does it mean when a mineral has a definite chemical composition?

4. What two elements are found in the rock-forming tetrahedron?

5. What is the overall charge for the tetrahedron mentioned above?

6. What is the chemical formula for olivine?

7. How is temperature important in mineral formation?

8. What are the six groups of non-silicates?

9. What is a ferromagnesian silicate?

10. Name three mineral properties that are used for mineral identification.

CHAPTER 3

IGNEOUS ROCKS AND IGNEOUS BODIES

In the previous chapter we talked about how magma/lava will crystallize to form igneous rocks. In this chapter we will learn about different rock textures associated with different cooling rates as well as how we classify igneous bodies. Igneous bodies are large exposures of igneous rocks. In the above photograph we see the easily the recognizable Half Dome from Yosemite National Park. At one time in the past, all the rocks you see in the photograph of Yosemite were cooling deep underground in a large body of magma. When magma is cooling beneath the surface of the Earth, we say it is cooling as an intrusive. Intrusive igneous bodies cool and crystallize within the Earth and this can be at any depth, as long as it is beneath the surface. When magma reaches the surface and becomes lava, the cooling and crystallizing rock is called an extrusive. Extrusive igneous rocks form on the surface of the Earth and intrusives form within the Earth. We can determine if a rock is intrusive or extrusive by the crystallization texture expressed in the rock.

Intrusive textures

Deep intrusive igneous rock textures are easy to spot because these rocks grow large crystals, and these textures are formed exclusively by the rate of cooling. Magma is hot and it will cool and solidify to form rocks. The slower this happens, the more time the tetrahedrals have to bond and grow. So as magma cools slowly, we will see larger and larger crystals being formed. When you can see and differentiate the minerals in a hand sample, this slow cooling texture is called **phaneritic**. When you see a rock that has phaneritic texture, you can now make the assumption that this rock sample cooled slowly. Figure 3.1 has many crystals that we can see with the naked eye, and these include K-spar, biotite, quartz and Na-plagioclase. This is an example of phaneritic texture.

We said that deep intrusive rocks will grow these large crystals (phaneritic) and now we need to talk about why. We said earlier that intrusive rocks could be at any depth below the surface. The deeper the cooling material, the slower it will cool. So rocks at great depth will have larger crystals.

Figure 3.1 **This is an example of phaneritic texture.**

As we approach the surface, however, the temperature drops. We are still considered to be an intrusive texture but magma that is cooling near the cold surface will crystallize much faster than its deep cooling counterpart. This accelerated cooling produces much smaller crystals. Generally we consider these to be so small that you cannot identify individual crystals with the naked eye. This rapid cooling produces an **aphanitic** texture. In figure 3.2, you can see that it is basically one uniform color. You cannot pick out the individual crystals in this hand sample. These are found in igneous bodies that cool near the surface, and will talk about the specific igneous bodies

Figure 3.2 **Example of aphanitic texture.**

later in this chapter. So deep intrusives are phaneritic, and shallow intrusives are aphanitic. But what happens when a rock starts to crystallize at depth and then is suddenly thrust up to a more shallow cooler crustal level?

This scenario of initial crystallization at depth followed by shallow rapid cooling will produce a mixture of textures. This will be expressed as larger crystals (phenocrysts) suspended in a matrix (ground mass) of finer crystals. We call this **porphyritic** texture. This indicates a mixed or more complex cooling history as compared to phaneritic and aphanitic. This type of texture is commonly found associated with volcanic systems that provide the mechanisms for rapid exhumation. In figure 3.3 you can see the light colored phenocrysts suspended in the light grey colored ground mass or matrix.

These three textures represent intrusive cooling rates. Later in this chapter we will associate textures with specific intrusive igneous bodies. This will allow you to make assumptions about the cooling history of the igneous body based on the textures present.

Figure 3.3 **Example of porphyritic texture.**

Extrusive textures

Extrusive textures are textures that form from the crystallization of lava. All of these rocks will crystallize on the surface. As you can probably guess, these textures will be formed from the rapid cooling of molten material. This material is generally considered to have been ejected by some volcanic process. Our first extrusive texture is **glassy**. This forms when the lava is ejected from a volcano and it cools extremely rapidly, often while still aloft. This rapid cooling is so fast that the crystal lattice does not have time to form properly. This produces a rock called obsidian. You can

see the glasslike appearance in figure 3.4. This rock has the correct chemical composition that would have allowed it to grow numerous individual minerals, but due to the improperly formed lattice, a glass is produced.

Another rapid cooling example is called **vesicular** texture. This is generated when the cooling lava has a large volume of dissolved gases called volatiles. As the molten material approaches the surface, the pressure on the

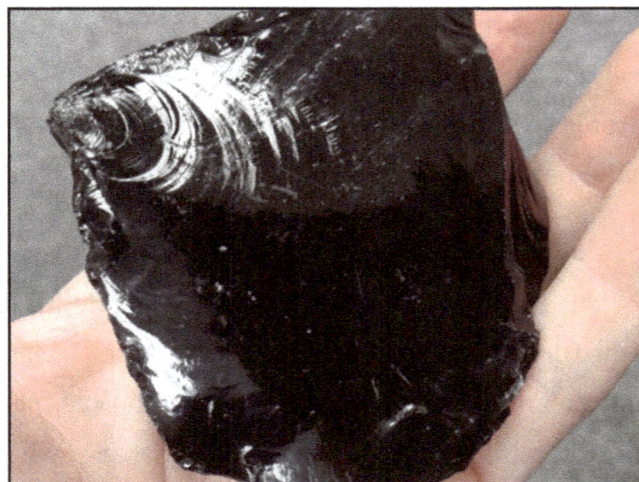

Figure 3.4 **Glassy texture of obsidian.**

material drops, and this will allow for the dissolved gases to be released. This outgassing causes the hardening material to be filled with holes (vesicles). All of these holes result in the rock having a low density. Low enough that it will float in water. For example, think about taking a can of diet soda and shaking it vigorously. Once you have it properly shaken, hand it to a friend and have them open it. The "foam" that is rapidly released represents the CO_2 outgassing from the diet soda. If this foam could harden, it would form a rock with vesicular texture.

Figure 3.5 shows an example of vesicular texture. You can see all of the void spaces that were produced by the outgassing once the pressure on the molten material dropped. This is what gives this material its low density. Figure 3.6 is an enlargement of figure 3.5. In this image you can see what might be described as stretched filaments. This indicates that the material was flowing and moving as it was hardening. This tells us that the sample is from a pyroclastic flow, but more on those in later chapters.

The third type of extrusive texture is actually not related to cooling rates at all. It is a texture that is made from smaller pieces of rock that were blasted from a volcano and were later welded together. This secondary texture is called tuff or welded tuff (figure 3.7). You will have a sample of this

Figure 3.5 **A rock with vesicular texture.**

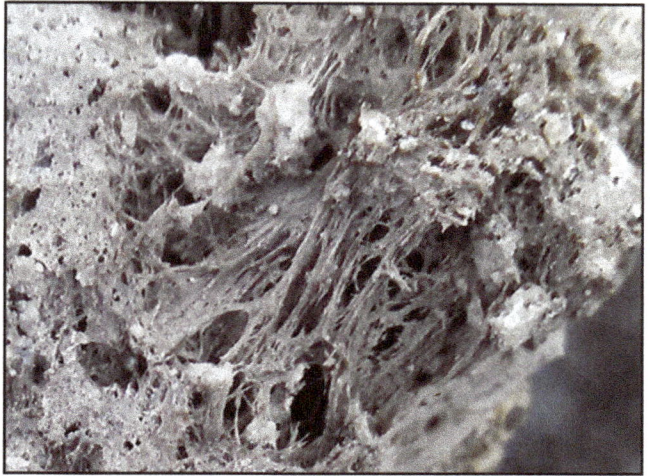

Figure 3.6 **Enlarged image of vesicular texture.**

Figure 3.7 **Tuff or welded tuff texture.**

texture in lab. Make sure you observe the smaller clasts from other volcanic rocks within this welded material.

Igneous bodies and their textures

Now that we understand that the cooling rate of igneous rocks controls the observable texture, we need to assign terminology to the igneous bodies. This terminology is based on the size, shape, and their affiliation with their host rocks. As we talk about these various igneous bodies, I want you to be thinking about their cooling rates and associated cooling textures. All of these igneous bodies will have one type of intrusive texture or another, suggestion that all of these igneous bodies will cool underground. We will talk more about extrusive igneous rocks in the chapter that covers volcanoes. By definition, intrusive igneous rocks are hosted by preexisting **country rock.** These different bodies will penetrate and cut through this country rock and cool at different rates that will ultimately correspond to the temperature of the country rock. Figure 3.8 shows several of the features we will be talking about and their relationship to the country rock.

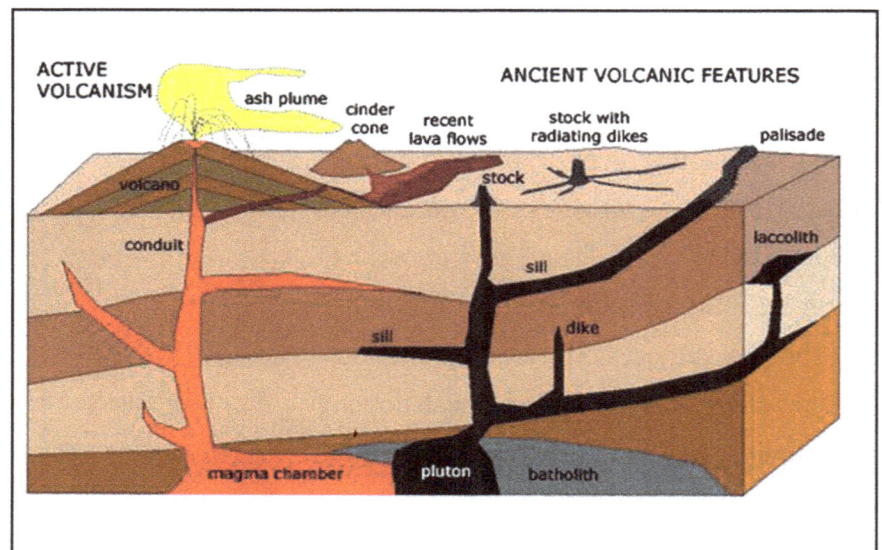

Figure 3.8 Rock formation diagram.

Courtesy of USGS.

Plutons

The first igneous body that we want to talk about is also the largest. These are deep intrusions called **plutons**. They form at considerable depths beneath the surface when rising magma becomes trapped in the cooler crust. Due to the great depths, the host or country rocks are much warmer than rocks near the surface. This allows for the mass to cool at a slow rate, and this will produce a phaneritic texture. Plutons are divided into two distinct categories that are based on the size of their surface exposure. I do want to point out that it is surface exposure, and this could mean there is a large amount of material below the surface that is just not exposed at present. With that being said plutons that have exposures larger than 100 km² are called **batholiths** and plutons that are smaller than 100 km² are called **laccoliths.**

Most of us are familiar with the nearby laccolith, Stone Mountain, Georgia. It can be seen rising up from the surrounding area of Atlanta as you drive

Figure 3.9 Stone Mountain, Georgia.

George Allen Penton/Shutterstock.com

down 85 South. The difference between laccolith and batholith can be easily understood when you compare the size of Stone Mountain, GA (figure 3.9) with another well-known batholith (figure 3.10).

Dikes and sills

Shallow intrusive igneous rocks are generally considered to form at depths around 2 km. This shallow depth means that the country rocks here are much cooler than the ones the plutons are cooling in. These shallow intrusives will cool rapidly, and this will generate an aphanitic texture. There are two classifications for shallow intrusive igneous rocks, and it is based on the intrusive rocks' relationship with the country rock. The classifications are **dike** and **sill**. What separates dikes from sills is the manner in which these two cut or intrude the country rock. Dikes will cut across preexisting strata, layers, or foliation in the country rock. In figure 3.11, you can see a dike cutting across the image from lower left to top right (GPS unit for scale). have highlighted the layers of the country rock in red. In this case, the layers that are being cut by the dike are metamorphic foliation. We will learn more about foliation in upcoming chapters. It is important

Figure 3.10 Yosemite National Park is a surface exposure of one large batholith. All the rocks you see in this image are associated with this one intrusive feature. Maybe this will help to illustrate the difference in scope between batholiths and laccoliths.

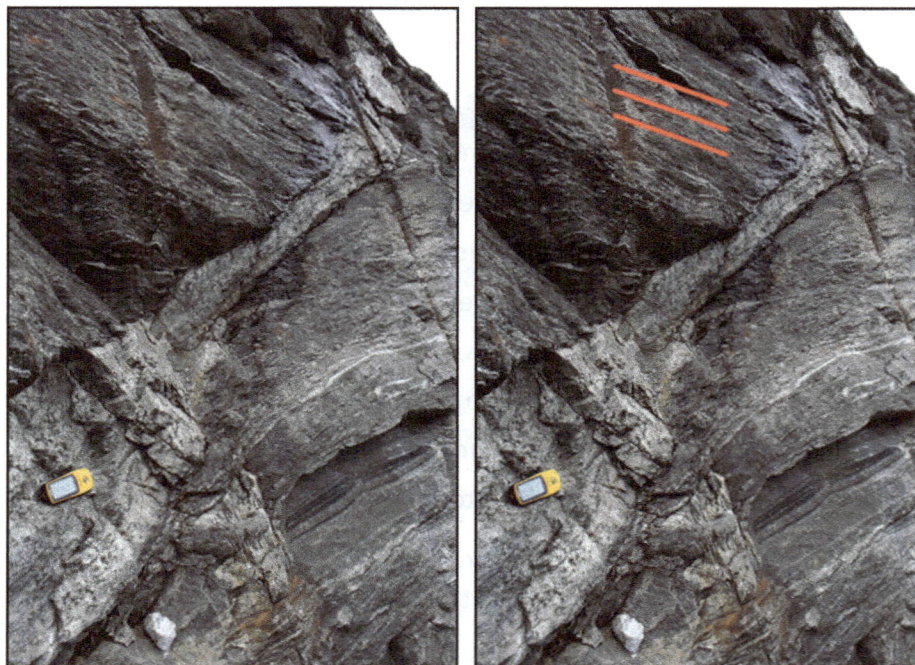

Figure 3.11 a and b

Figure 3.12 DIKE.

to note that dikes will cross cut features, and it does not matter if those are strata (sedimentary features) or foliations (metamorphic features).

When you view the dike in its entirety, you can clearly see that it is cutting the preexisting layering in the country rock. Dikes cut layers, and that is easy to see and understand (figure 3.12). What about when the rock does not have easily defined layers, is it still a dike? Yes. Dikes can cut across plutons that do not express any foliation. There are no layers, however, the dike is still considered to be cutting across that feature.

In figure 3.13, you will see several different features. In the center, there is a small thin lighter colored feature, and this is a dike. You can clearly see it cutting across the host rock even though preexisting layers are not readily apparent to the untrained eye. In addition to the dike, you can see a dark colored feature that is cut by the small dike but also appears to cut the host rock. This is in fact not another dike, but rather it is a **xenolith**. A xenolith is a rock fragment which becomes enveloped in a larger rock during the crystallization of the en-capsulating rock. In this case, the encapsulating rock is a pluton and the darker band of material is a fragment of the original host rock. Both the pluton and the xenolith were later cut by the small dike we discussed earlier. In class we will see several exam-ples of xenoliths. By defi-nition, the xenoliths will

Figure 3.13 **Several features evident.**

Figure 3.14 **Sill at Top of Hill.**

Figure 3.15 **Sill is highlighted.**

always be older than the encapsulating intrusive rocks. We will use this fact to our advantage later in the semester when we are constructing geologic sequence of events.

Sills are very similar to dikes. Sills fill in between preexisting layers of host rock. They do not cut across layers, but rather they are confined by them. In figure 3.14, we can see an exposed ridge on the top of a hill. This ridge is actually igneous material that intruded parallel with the surrounding layers producing a sill. In this case, the igneous material is more resistant to weathering and erosion than the host rocks, and that is why it is forming a ridge. This sill is parallel with the surrounding strata. If it had cut across this strata at any point, it would stop being a sill and become a dike. Remember, dikes and sills are very similar. They are only differentiated by how they intrude.

Figure 3.16 **Devil's Tower, Wyoming.**

Scott Sanders/Shutterstock.com

Volcanic neck

The last igneous body that we will talk about in this chapter is a **volcanic neck**. Just as the name implies, these igneous bodies will form in the neck of a volcano. As hot magma rises up the conduit of a volcano, it can often stall and cool in place. This forms a volcanic neck or volcanic plug. This igneous material will preserve the shape of the volcanic neck. As time passes, the less resistant volcanic material that had formed the cone will be eroded and the neck is left exposed. Over time this too will be eroded, but for a brief period of geologic time, it will stand as evidence of past volcanic activity.

Figure 3.16 is a volcanic neck called Devil's Tower. It is a popular visitor attraction in Wyoming, and its origins are woven into the regional folk lore. One of the more prevalent stories about this igneous feature is that a great bear tried to climb the mountain and this caused all the "grooves" you see along the sides. Rather than bear claw marks, this is a feature called columnar jointing. We will talk about how this forms in class, and we will look at some other famous examples of this cooling feature.

Questions

1. What are the three types of intrusive rock textures discussed in this chapter?

2. How does the cooling rate determine the texture?

3. What does it indicate about the rock if there are stretched filaments within vesicular texture.?

4. What is a county rock?

5. What is the difference between batholiths and laccoliths?

6. What is the difference between dikes and sills?

7. Define xenolith.

8. How are volcanic necks formed?

Additional topics

Please research the term **columnar jointing** and understand how this feature is formed.

CHAPTER 4

WEATHERING
PHYSICAL AND CHEMICAL WEATHERING

In this chapter, we will talk about various processes that are hard at work trying to destroy the rocks that are exposed to surface conditions. In the last chapter we talked about igneous bodies, laccolith, and batholiths for example, and when these igneous bodies are exposed, the weathering process begins. The weathering process will attack all rocks exposed at the surface. To this point, we have only talked about igneous rocks, but it will also affect sedimentary and metamorphic rocks (more about these in upcoming chapters).

What is the weathering process?

The weathering process is simply the breakdown of a preexisting rock. Of the three rock types (igneous, sedimentary, metamorphic), this breakdown can and does happen at different rates. Some rock types are much more resistant than others. Take your granite counter tops as an example. Those are igneous rocks and are very resistant to breaking down, if not they would make very poor counter tops. In addition to rock type, climate will also play a role in determining the rate of weathering. As we look at specific types of weathering in this chapter, think about how various climates would affect their rates.

We are going to break the weathering process down into two different categories. We are going to think of it as mechanical and chemical weathering. Mechanical weathering is simply the physical disintegration of a rock. This will take larger rocks and break them down into smaller pieces. To help set the stage for next chapter, it is these small pieces that will become the material used to form sedimentary rocks. Chemical weathering will go a step beyond the physical breakdown and actually chemically dissolve the rock.

Physical weathering

There are four types of physical weathering that we will talk about in class. All four types will use a different mechanism to physically break down rock material. Even with the mechanisms being different, the end process is the same—*the mechanical disintegration of rock material*. We are making the rocks smaller.

1. Biological activity

The mechanical weathering category that is defined as biological activity refers to the physical break down of rock material caused by any form of a living action. This activity can include plant growth, animal activity, and interference caused by humans. In figure 4.1, you see how the simple growth of plant roots can fracture granite. When the plant began to grow, the small roots took advantage of small micro fractures in the rock. As the roots grew, they were able to widen and open the cracks and eventually break the rock into smaller pieces. Also in the image you will see a small metal cylinder in the lower left

Figure 4.1 Tree roots actively exploiting cracks in the country rock.

of the image. This is a movement detection device. This will measure and detect any movement of the rock due to more advanced weathering. There is one force that we must always remember when talking about weathering. That is gravity. Yes, the actual root is fracturing the rock, but gravity will pull the rock apart. Gravity will aid in this weathering process.

Biological activity can come in an array of different shapes and sizes (figure 4.2). It can range from worm burrows to large-scale animal activity. In addition to tree roots wreaking havoc, smaller plants also play a large role. When you see mosses and lichens growing on exposed surfaces, just realize that they are also causing biological break down of the rock face.

Figure 4.2 Variety of biological activity.

Figure 4.3 shows grooves (claw marks) from penguins repeatedly walking along the same path. This simple wearing away of the rock is classified as biological weathering. It is exposing a greater surface area of the rock, and this will actually serve to speed up the weathering process.

2. Abrasion

The mechanical weathering category that is defined as abrasion refers to the physical break down of rock material caused by any form of a particle to particle interaction. Put simply, abrasion is caused when rocks bump into one another in rivers, streams, glaciers, deserts, and shorelines. This particle interaction will decrease the overall size of the particles involved. In addition to reduced particle size, abrasion will also work to abrade channels. When particles are carried by wind, water, or ice they can behave similar to sandpaper and wear away at the rocks

Figure 4.3 **Penguin claw marks are another type of biological weathering.**

below. This process will cause river channels and glacial features. The glacier in figure 4.4 is slowing moving down the mountain side. In addition to the weight of the ice, the bottom of the glacier is carrying rock material that it has picked up on its journey down the mountainside. This rock material will

Figure 4.4 **Glacier.**

slowly carve out/abrade the rock material beneath leaving a glacial valley behind. This process is also observed in rivers and in wind-blown regions.

In figure 4.5, the Rio Grande River is slowly cutting through many layers of volcanic material that is exposed near Taos, New Mexico. The river will carry small abrasive particles like a conveyer belt and continue to cut deeper into the canyon. This is also re-sponsible for some of the more stunning National Parks. The Grand Canyon and Zion Canyon are examples of this process.

3. Freeze thaw/Frost action

The mechanical weathering cat-egory that is defined as freeze thaw refers to the physical break down of rock material caused by

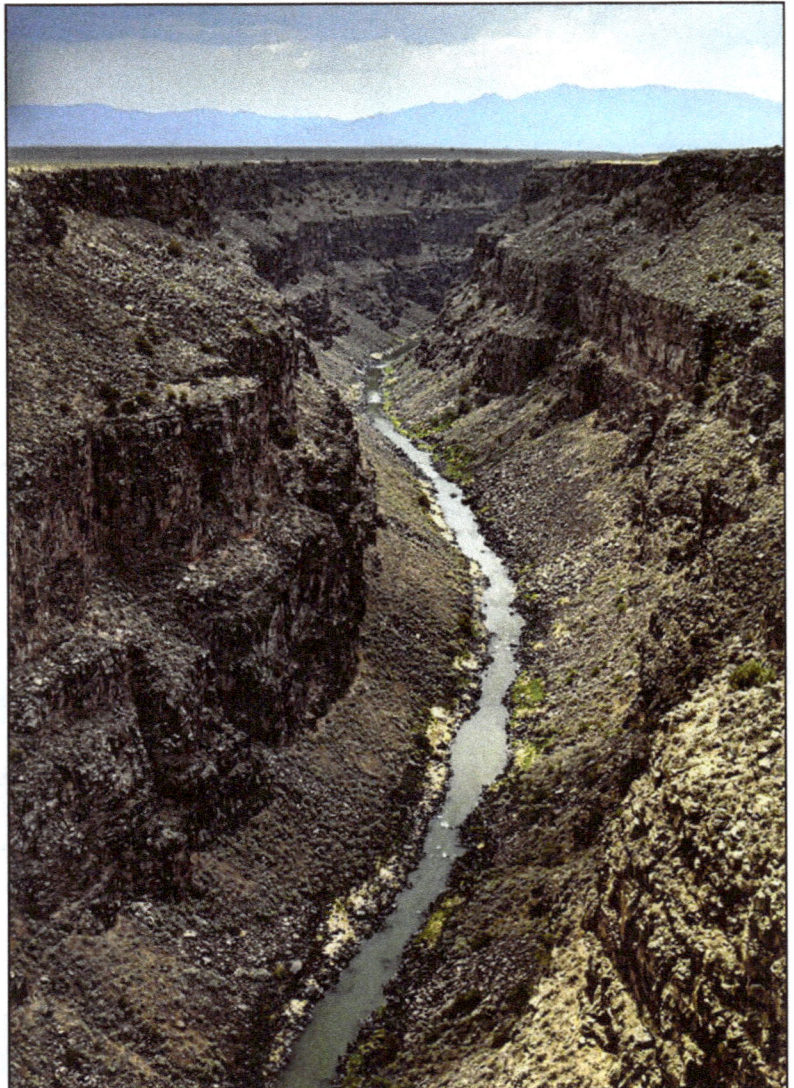

Figure 4.5 **Rio Grande River.**

the force of water freezing and thawing over and over in small cracks within the rock. This pro-cess is simple yet effective. When it rains, water will fill cracks found in the surface of the rock. As the temperature drops at night, the water will freeze and expand. This expanding ice will apply enough force to the preexisting crack so that the crack will slightly propagate deeper into the rock. The next day, when the ice melts, the water will now penetrate deeper into the newly widened crack. When this cycle happens repeatedly, large boulders can be splintered.

This process will only be found in climates where the temperature will regularly be above freezing during the day and below freezing at night. This particular climate is common in higher

Figure 4.6 **Rock runs on the Falkland Islands.**

latitudes. In addition to frost action, these environments can have multiple freeze thaw processes occurring simultaneously. These include soil fluction, frost heaving, frost sorting, and washing. These process can combine to produce interesting and unusual rock formations. One example would be the Rock Runs found on the Falkland Islands of the east coast of Argentinia (figure 4.6). These rocks are "flowing" down the hill due to the action of freeze thaw. As the water freezes, it slightly lifts the rocks off the surface. Due to the pull of gravity and the slope of the terrain, the rocks slide down hill a few centimenters at a time. The ice melts and drops the rocks a few centimeters farther down the hill. As this cycle repeats, rocks appear to flow down the hill side.

4. Pressure release

The mechanical weathering category that is defined as pressure release refers to the physical break down of rock material caused by the release of pressure on a rock body. This pressure is generated by the weight of the column of rock that is found on top of a buried rock body. Imagine a rock body like a laccolith or a batholith that has cooled underground. This intrusive body will be under a steady pressure, and the rock will respond to this over burden stress. When the over burden is released, the rock body will expand and enlarge. This expansion will cause the outer areas of the rock to break off in layers. In class we will look at some examples of this exfoliating type of weathering. The resulting weathered dome can also be referred to as an exfoliation dome. This type of weathering is easily found at Stone Mountain, Georgia.

Another well-known example can be found in Yosemite National Park. In figure 4.7 you will see Half Dome. Its unique and iconic shape was carved by glaciers during the last ice age. In addition to glacier/abrasive weathering, we can also see exfoliation. In the center of the image

Figure 4.7 Half Dome, Yosemite National Park.

you will see several layers that appear to be peeling off the side of the rock face. This is result of pressure release. The rock body is expanding, and as a result the outer layers of rock are lost.

Chemical weathering

Mechanical weathering makes smaller pieces by physically breaking down rock material, thereby, increasing the *surface area* available for chemical weathering. Unlike physical weathering that physically breaks the rock down into smaller pieces of the parent material, chemical weathering will change/alter the parent material. The agents of chemical weathering that we will be covering in this course are water (H_2O) and weak acids. This type of weathering is most commonly found in warm wet climates. Tropical regions will have higher rates of chemical weathering rather than physical weathering. The extensive soils that are found within areas adjacent to the equatorial regions are primarily due to chemical weathering.

There are four examples of chemical weathering that we are going to talk about in class. The four equations will be listed below. Details will be discussed in class.

1. ***Dissolution***—rock material dissolves leaving no residue, everything ends up in solution.

Solution

$$CaSO_4 * 2H_2O ---> Ca^{+2} + SO_4^{-2} + 2H_2O$$

2. ***Carbonatization***—CO_2 is agent that produces carbonic acid.

$$H_2O + CO_2 ---> H_2CO_3$$

      Rain   in atmosphere   carbonic acid

step 2

$$H_2CO_3 + CaCO_3 ---> Ca^{+2} + 2HCO_3$$

        Calcite   Solution

3. ***Hydration ("Hydrolysis")***—adds water and hydrates the mineral

$$2KAlSi_3O_8 + H_2O + 2H^+ ---> 2K^+ + Al_2Si_2O_5(OH)_4 + 4SiO_2$$

   orthoclase   kaolinite   solution

                                    (clay mineral)

4. ***Oxidation***—O_2 is the agent; oxygen changes valence ("ionic") state

$$6H_2O + 2Fe_2SiO_4 + O_2 ---> 4Fe(OH)_3 + 2SiO_2$$

          olivine   limonite

All four of the above chemical weathering reactions are active in the southeast. Due to the warm climate and abundant rain fall, we find that most of our rocks exposed at the surfaces have either turned to clay, oxidized, or both. Figure 4.8 shows the effects of oxidation. You can see the layer of orange/red on the surafce of this outcrop. Once the oxidized layer has been removed, the bright colors of the originial clay can be seen. It is this oxidation process that creates the famous Alabama red dirt that is nearly impossible to wash out of your clothes. The next time you drive down a dirt road, notice the color. If it is deep orange/red, then it is probably due to oxidation.

Figure 4.8 **Effects of oxidation.**

Differential weathering

As we have discussed earlier, climate can affect the rate of weathering. In addition to climate, the rock type will also play a role in how fast that particular rock will weather. Remember the Mohs hardness scale, rocks of a higher degree of hardness will weather slower than the ones below them on the scale. This varying rate in weathering is called differential weathering. There will even be situations where one outcrop will have portions of rock that are more resistant than other sections. Large exposures of rock, especially sedimentary rocks, will have a high degree of heterogeneity across the exposure.

When you have this variance, you will see different weathering rates.

In figure 4.9, this undergraduate geology student is pointing out a case of differential weathering. She is pointing to small circular holes that are in the surface of a limestone. At first glance, it may appear as if these were man made.

Figure 4.9 **Differential weathering can cause circular holes in limestone.**

Figure 4.10 **Differential weathering gives rock face a jagged appearance.**

Figure 4.11 **Example of differential weathering in the Grand Canyon.**

They are completely natural. Spherical and circular weathering is common and we will discuss the mechanisms behind this weathering shape in class. This particular outcrop is on a mountain top in northeast Georgia, and it is exposed to physical and chemical weathering processes, and it is a combination of both that produces these interesting circular formations.

Another example of differential weathering can be seen in figure 4.10 from the Grand Canyon National Park. You can see what looks like pieces of rock protruding out from the surface of the outcrop. This would be an adequate description. The material that seems to be sticking out from the rock is actually weathering out of the solid material. This means that this material is harder than the surrounding sedimentary rock. As the rock surface weathers, the softer material surrounding these harder nodules will weather away. These resistant nodules (mostly quartz) will weather slower and can even eventually weather out of the rock completely. This example of differential weathering is what gives this rock face a jagged appearance.

Questions

1. What are the two types of weathering?

2. Explain how abrasion works to weather rocks.

3. Explain freeze thaw/frost action.

4. Describe pressure release.

5. What are the four types of chemical wreathing?

6. What is differential weathering?

7. What chemical weathering reaction depends on carbon dioxide in the atmosphere?

8. What chemical weathering reaction depends on free oxygen in the atmosphere?

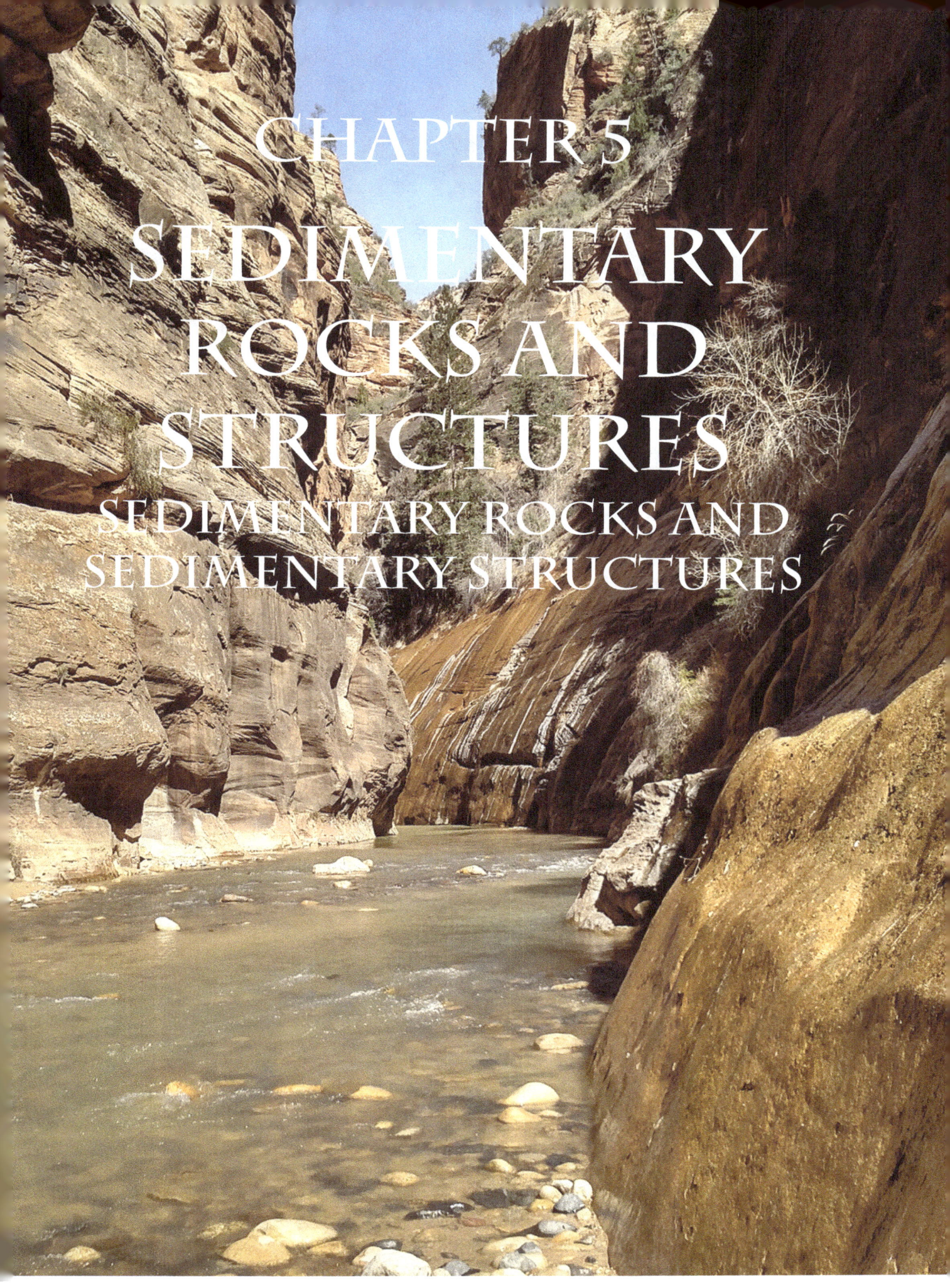

CHAPTER 5

SEDIMENTARY ROCKS AND STRUCTURES

SEDIMENTARY ROCKS AND SEDIMENTARY STRUCTURES

From weathering to sedimentary rocks

In the last chapter we talked about the various ways physical and chemical processes can break down and alter rock material. In this chapter, we are going to see how that weathered material will come together and form sedimentary rocks. In the weathering chapter, we were primarily talking about the weathering of igneous rocks, but weathering will affect all three rock types (igneous, sedimentary, and metamorphic). So the sediments that we will be discussing could have been generated from the weathering of any of the above rock types. Geologists have several observational techniques which they can use to help them tell from what material the rock has come from. We call this sourcing rock the **parent rock**. In many cases it can be sedimentary rocks that are weathering to form new sedimentary rocks.

Once we have weathered material from a parent rock, it is often transported, and can be transported to great distances from the parent material. In some cases, the weathered material can form rocks that stay close to home, but when sediments travel they take on characteristics that help us determine just how far they have traveled. The first part of this chapter will focus on what happens to these sediments as they travel. Let's just think about a hypothetical situation for a moment. Imagine you are a piece of granite high up in the mountains. You are weathered into sediment and you begin to travel. How far will you go? How far can you go? To think about answering these questions we need to first realize where sediment can and will accumulate. Any place that sediment can accumulate is called a depositional area. There are several environments that are ideal for sediment accumulate; oceans, lakes, deserts, and beaches just to name a few. In addition to understanding sediment transportation in this chapter, we will also see what the rocks can tell us about their **environment of deposition**.

We have two photographs showing beaches (figures 5.1 and 5.2). One photograph is taken on the east coast of the Atlantic Ocean in Scotland, and the other is taken on the west coast of the Atlantic Ocean in Maine. As you can see in the photographs, the beaches look very different from each other. One of the first observations that can be easily made is the difference in sediment size. When we refer to the size of sediments, we are talking about grain size. When we classify sediment, we use its grain size. So you can see that the size of the sediment or grains is very important in classifying the sediment and the sedimentary rock that forms from it. If weathering is the breaking down of grains into smaller grains, it is clear by the photographs that the rocks in Maine still have a considerable amount of weathering to undergo to look like the beaches of northern Scotland. The amount of time that sediments spend being transported will have a direct impact on several characteristics that we can observe about sediments.

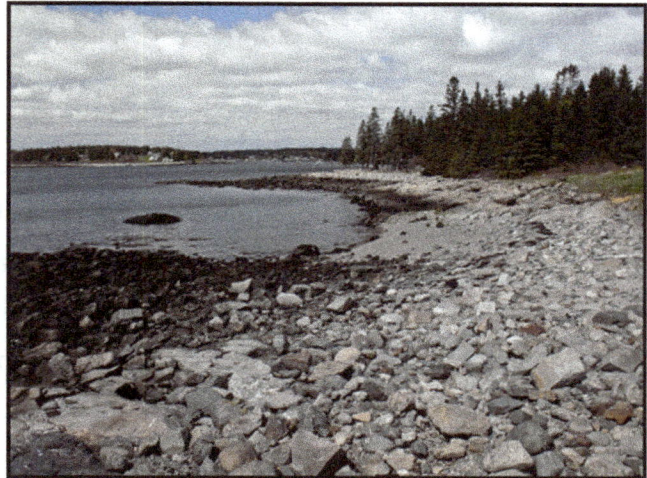

Figure 5.1 **Atlantic Ocean in Scotland.**

Figure 5.2 **Atlantic Ocean in Maine.**

All sediment must take the same path

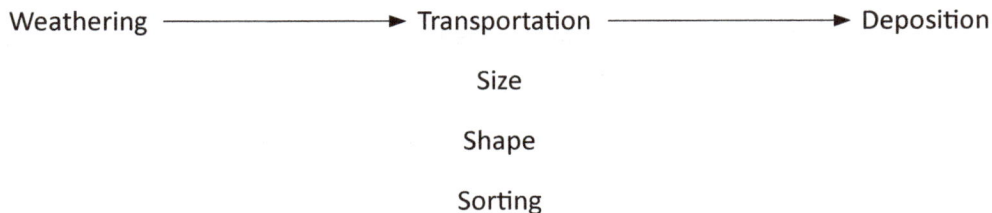

Weathering ⟶ Transportation ⟶ Deposition

Size

Shape

Sorting

Sediments can be transported in two broad categories. Sedimentary transportation is the way sediment will reach the site of deposition and begin the transition from sediments to sedimentary rocks. The first category is called the **suspended load**. This includes all the sediment that is small enough to be suspended and dispersed throughout the flow, and in addition to suspension, this also includes material that we will consider to be dissolved into the water column (material from chemical weathering). This material will flow with the speed of the current and the rate of flow will also control when this material settles out of the water column. When fine-grained material is floating or suspended in the column, it will eventually drain out of the water column once the velocity of the flow has slowed. The velocity needed will depend on the size of the particle suspended. The material that is dissolved into the water column will require more changes than just velocity. The correct chemical characteristics will need to be met in order to precipitate this material.

The second category that pertains to sediment transportation is called the **bed load**. This contains material that is too large to be suspended and must move along the bottom. The size of particle that can be picked up and carried in the water column depends on water velocity, so the size of the material found in the bed load will also be depended on water flow velocity. The faster the river, the larger the particles that are found in the bed load. For example, a river that is often used for white water rafting will have a very high velocity and that flow can easily move particles that could be fist sized or larger. A quiet flowing river, one that is more suited for floating or tubing, will struggle to pick up particles larger than fine sand off the bed. Both of these transportation mechanisms are dependent on velocity.

The above two categories are specifically talking about particle transportation in a liquid media. Are there other fluids that can transport sediments? In addition to water, wind and ice are also excellent methods to transport sediment. Wind and water have similarities in the method of transportation. In addition to suspended loads, you will also see **saltation** and **traction**. We will talk about these two methods in class. Saltation is basically bouncing the particle along on its path and traction is rolling usually at the bottom of the column. It is this particle to particle interaction that will continue to physically break the particles down even smaller. This physical breakdown will gradually cause the particles to become more spherically as they are transported. This rounding effect will help us gauge how long the particle has been in the transportation system.

When particles first enter the transportation system, they are generally close to the source or parent rock material, and under close observation, they will be very angular in nature. Angular means that the particles/grains will have sharp edges. The grains have not been abraded yet by the particle to particle interaction generated by saltation or traction. This angularity indicates that the particles have only been transported a short distance or short amount of time. As this distance/time increases, the grains will become more and more rounded. We generally divided the sediments up into three classifications. Angular, rounded, well-rounded with angular indicating the grains first entering the system and well-rounded indicating grains that have spent the most time being transported. So when a geologist examines sediments, the degree of rounding can indicate how long and also how far the grains have been traveling.

Another useful tool for determining duration of transportation is the degree of sorting. **Sorting** is the grouping together of grains by similar sizes. When you go to the beach for spring break and you take that first walk along the beach, I want you to think about the degree of sorting. If you find yourself on the white beaches of the Alabama or Florida gulf coast, take note that as far as your eye can see, the sand grains making up the beach are all roughly the same size. There are no large boulders or even large stones found in the mix. If there were, it would make beach volleyball much more interesting. This is considered to be very well sorted. The classification systems ranges from very poorly sorted, poorly sorted, moderately sorted, well sorted, and very well sorted (*Not all geologists are creative as others*). This, like rounding, serves to indicate distance transported. Very poorly sorted being new to the system with very well sorted at the opposite end.

NOTE: Of the three media we mentioned earlier, wind and water are excellent at sorting particles. Ice, on the other hand, is a terrible sorter.

Now all this talk about rounding and sorting leads us to the most important topic of all, **grain size.** Grain size is one of the most important classification techniques that a geologist will use. It not only classifies the sediment, it is also critical in naming the rock that is formed from the varying sizes of sediments.

In table 5.1 you see four different grain size classifications: pebble, sand, silt, and clay. We use these four terms to reference sediment that falls within predetermined size parameters. The word sand can be used for any particle that is between 1/16 and 2 mm. This is a broad

Table 5.1 Grain Size Classifications

pebble	>2 mm	conglomerate
sand	1/16–2 mm	sandstone
silt	1/256–1/16 mm	siltstone
clay	<1/256 mm	shales

category when you think about it. If geologists have to classify sediments on scales smaller than this, it would be a difficult task. So for this class, any particle that falls within that range we will call sand. With that being said, it is common in geology that there are times that we do need to subdivide the category of sand even further. We are not going to get into that, but I just wanted you to be aware that it is something that we can do when needed. So as with sand, if any sediment falls within the above size ranges, we can use the collective term. It is the collective term that we use to name the rock. So if the rock contains particles that are between 1/256 and 1/16 mm, we would be able to refer to the rock as a siltstone. Now that you see how the rock names are related to size, you will work with several of these sedimentary rocks in lab.

Using the finger as in figure 5.3 for scale, we can see that most of particles in this picture are smaller than 2 mm. There are a few that will be larger, but the majority will be sand-sized

Figure 5.3 Sand particles are smaller than 2mm (finger shown to show scale).

particles. When this turns into a rock, it will be called sandstone. The name comes from the size of the particles. In this next example (figure 5.4), we see sediments that have already become a rock. We have a camera lens cap for scale, and we can see that there are particles/clasts in there that are well over 2 mm. This rock would fall into the category of 2 mm or larger. It would be easy to just say conglomerate, however, we do have two subcategories within the classification of conglomerate. These two divisions are conglomerate and breccia. The way we decided which term to use depended on how the roundness of the large clasts appeared. If they are more rounded, we call

Figure 5.4 **Rock with camera lens cap to show scale. Are the large clasts rounded (conglomerate) or angular (breccia)?**

the rock a conglomerate. If the clasts are angular, we call the rock a breccia. So this difference is based on grain roundness alone. So think back to when we talked about roundness and you should be able to make predictions about conglomerates and breccias with regards to transportation. So take a look at figure 5.4 and see if you think the large clasts are rounded or angular. What do you think?

I actually think that I see both angular and rounded clasts in this example. The ones you have in lab will be much easier to determine, but I wanted you to see that the world of geology is not always cut and dry. So to answer this question, we need to look at all the clasts and obtain an overall average or estimate as to if we think they are rounded or angular. Also note the multiple rock types represented in this rock. There are coarse and fine-grained igneous rocks present. I hope you are now thinking back as to what that could mean for these igneous rocks. All of these are clues that a geologist will use to better understand the geology of the whole area.

In this last example, we looked at a rock instead of loose sediments like we looked at in the sand photo. In order for loose sediments to become rocks, they must undergo the **lithification** process. This process is a two-step process. First, the sediments must be compacted. We have to decrease the space between the particles so that it will be easier for them to be adhered together. This process of decreasing pore space is simply called **compaction**. Once the sediments have been compacted, they must be cemented together. This **cementation** step is what provides the coherency to form the actual rock. This cement is generally quartz or calcite that will precipitate between the grains. Once compaction and cementation are complete, the rock is lithified and it is now a sedimentary rock.

Sedimentary Structures

Primary structures in sedimentary rocks—form during deposition of sediments

The first primary structure that we need to talk about is *stratification.* Stratification is visible layers in a sedimentary rock; called "beds," "strata," etc. When you see layers in an outcrop, you can now draw the conclusion that you are looking at sedimentary rocks. Stratification forms by the accumulation of sediments, mostly due to gravity generated settling, into distinct layers. As sediments enter into a quiet environment, they will begin to settle. The environment needs to be quiet, or low energy or the particles will have a difficult time settling. As an example, imagine it's a hot summer day and you go into a restaurant and order a sweet tea. They bring out an ice cold glass, and with all the excitement and anticipation you can muster, you take a sip of what you find out to be an unsweet tea. After the initial horror has subsided, you add sugar to your glass as an attempt to sweeten it. When the particles are dropped into the glass, they promptly begin to settle. The particles will form a layer on the bottom of the glass. If you were to stir the glass with higher energy, the particles will go back into the water column and become suspend

due to the higher
energy. When you
stop stirring, the
energy drops and
the sugar settles
again. This is the
reason why sedi-

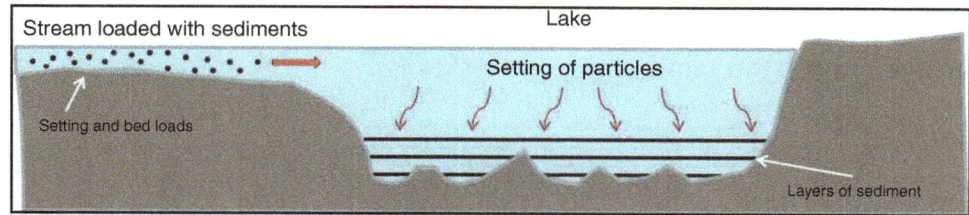

Figure 5.5 **Sedimentation settling process.**

mentary particles need a quiet environment to settle and form layers. The smaller the particles, the lower the energy state needs to be for them to settle. Sand will settle in an environment with much higher energy than a smaller particle, like clay for instance. Figure 5.5 illustrates how this process would happen in a lake.

Once these sediments become lithified and are exposed at the surface, we see them as layers preserved in the sedimentary rocks. What causes the division of these beds or layers are slight breaks or pauses in particle settling/deposition. After the slight pause, the deposition of sediments will continue, but this will now be the start of a new layer. In figure 5.6 you can see multiple beds of sandstone. This exposure is found right above the falls of Little River National

Figure 5.6 **Sand stone layers, Little River National Preserve, Alabama.**

Preserve near Fort Payne, AL. You can clearly see the layers of sand stone, and most of the layers are roughly the same size.

In this example, all the layers are made of similar rock types (sandstone). That is not always the case. In figure 5.7, we can see layers or beds that are composed of different rock types. These different rock types are caused by the fact that the particles making up the rocks are of different sizes. We have sand-sized particles making the sandstone (at the bottom of the image), and clay-sized particles making the shale layers that we see above the sand stone. Now that we understand how the energy of an environment can affect the size of the particles that will settle, we can begin to make assumptions about how an environment can change. Just a quick glance at this image, and

Figure 5.7 **Beds of different rock types.**

we can begin to develop a story about the past environment focusing on how its energy level was changing. It was becoming a lower energy environment because we see shale on top of higher energy sands. This combined with other clues help geologists unravel the mystery about the environmental conditions at the time of sedimentary deposition.

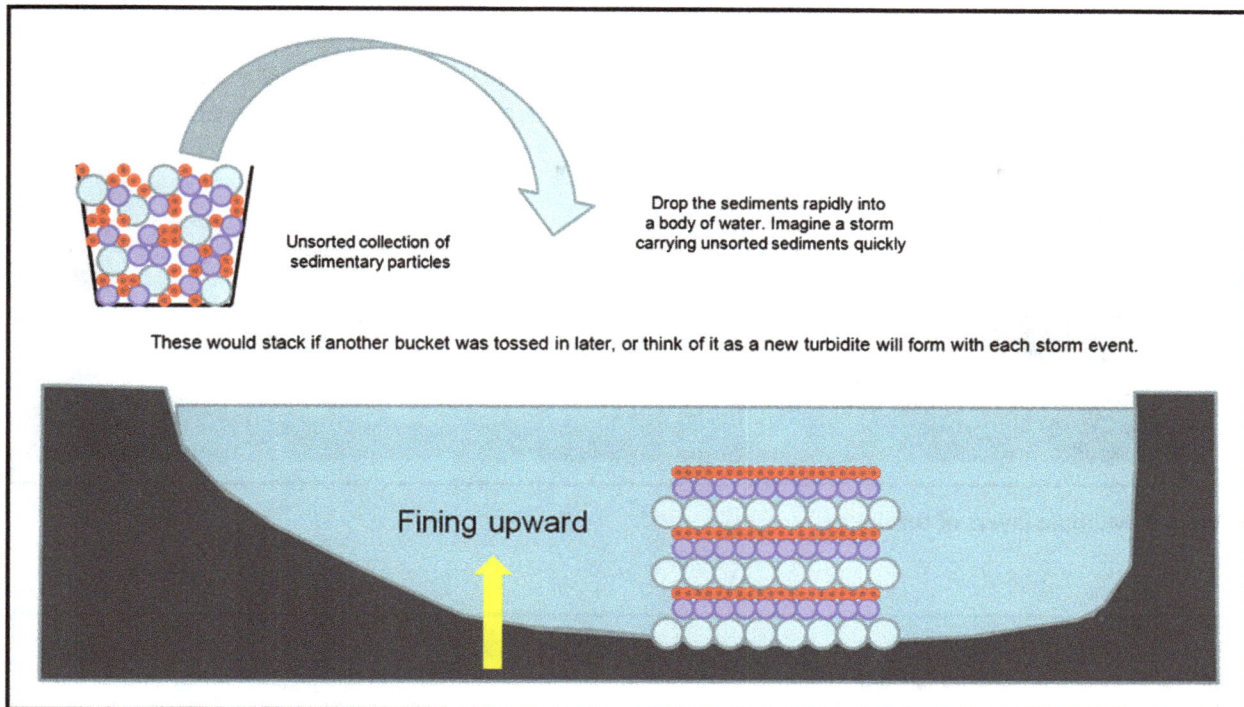

Figure 5.8 Fining upward sequence.

The second primary structure we need to talk about is a **turbidite**. This is a natural progression from the settling that we were just talking about in the above structure. We talked about how the energy level was partly responsible for the size of particle that would settle in that environment. That still holds true, but turbidites are a fining upward sedimentary particle size sequence. They are also called *graded beds* and can be very useful to geologists. A fining upward sequence is easy to recognize and the way it forms is very logical. Okay, imagine you take a bunch of different sized particles; sand, pebbles, silt, clay, and you throw all of these into a pool at once (figure 5.8). You can probably guess that the larger sized particles will fall to the bottom almost immediately, and as time passes, the smaller particles will settle on top of the larger. This will gradually build up a fining upward sequence. This will range from pebbles on the bottom to clay at the top.

Turbidites are very important to geologist for several reasons, but one important feature is that they show the direction of gravitational settling. Why would this be important to ask? It shows us which way was down when the turbidite was forming (figure 5.9). When we get to the structure section, we will see that it is not uncommon for rock layers to be upside down after they have been folded and faulted. By using turbidites we can tell if they are right side up or if in fact they are overturned. We will look at a few examples in class.

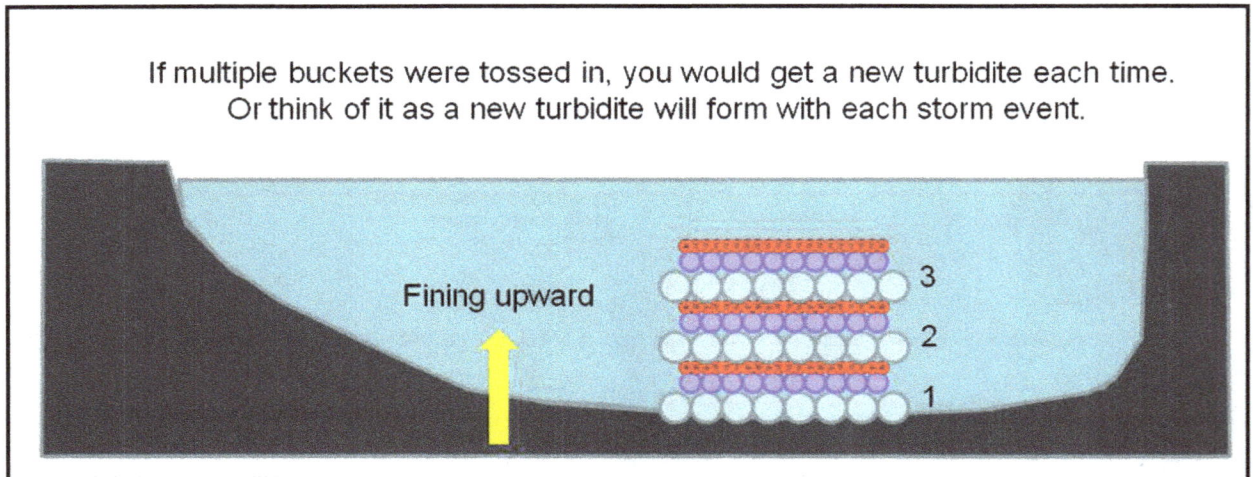

If multiple buckets were tossed in, you would get a new turbidite each time. Or think of it as a new turbidite will form with each storm event.

Fining upward

3
2
1

Figure 5.9 Multiple layers of turbidites.

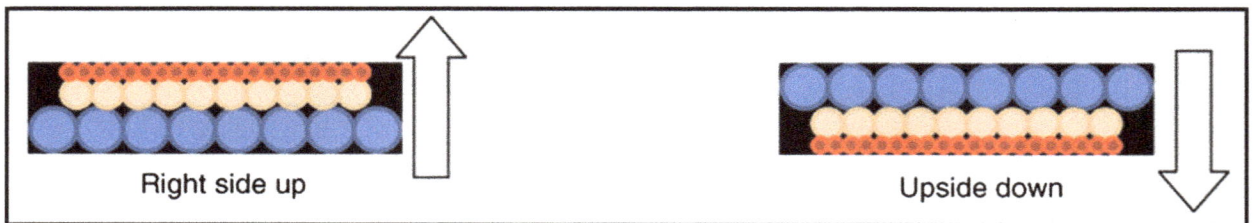

Right side up

Upside down

Figure 5.10 Turbidites showing orientation.

The third primary structure we need to talk about is ***ripple marks*** and associated ***cross beds***. Ripple marks are basically periodic waves within sand that are generated by flowing water and wind (figure 5.11). These ripple marks, or in larger scale dunes, help us understand the current direction of the wind or water that formed the ripple marks. Most of us are familiar with ripple marks. We have all seen them at the beach, and in the class power point we will look at some modern day examples. These ripples will have a specific type of asymmetry. It is this asymmetry that will indicate the past current direction.

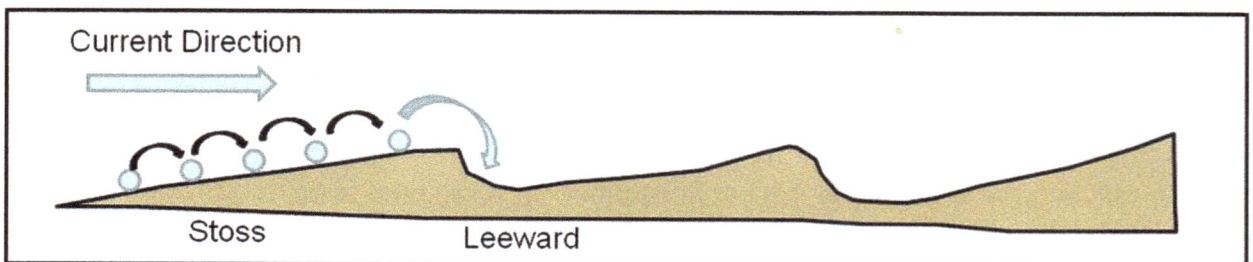

Current Direction

Stoss   Leeward

Figure 5.11 Ripple marks help understand the wind or water direction that formed them.

Figure 5.12 **Sandstone with preserved ripple marks.**

Current will move from the stoss side toward the leeward side. So if you can spot this asymmetry then you can accurately predict the past current direction. Figure 5.12 shows preserved ripple marks in a sandstone. You would be able to rub your hand across the rock and feel for this asymmetry. Samples like these are used to predict past paleocurrent directions. What do geologists do when this is not available? More often we do not get to see the top of the rocks to be able to check for this asymmetry. When this is the case we have to rely on cross beds. Cross beds are the expression of ripple marks in cross section view.

Cross beds will also have a very distinct asymmetry. They will intersect the top of the dune/ wave in a manner that is near perpendicular and then they will become tangential toward the leeward side. Since the tangential portion is directed to the leeward side, this can be used to determine current direction. In figure 5.13, we can see that all the cross beds are becoming

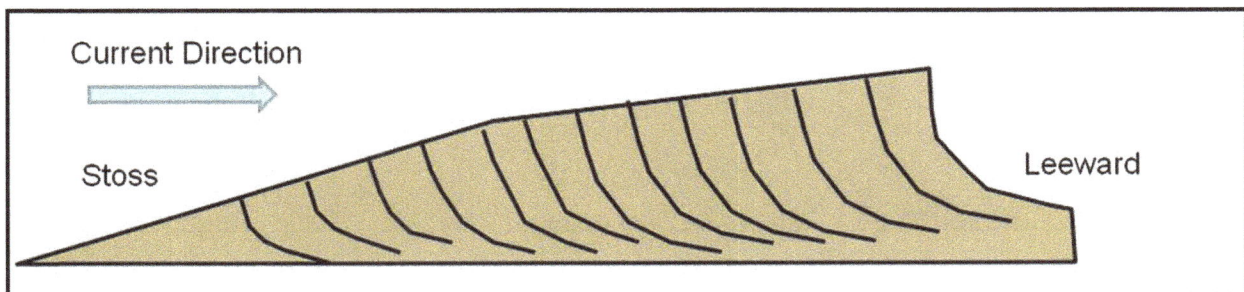

Figure 5.13 **Cross beds have distinct asymmetry.**

tangential at the bottom and are pointing toward the leeside, this indicates current direction. Let's take a look at some real word examples. Also, we will cover this material extensively in class. If you are having difficulty in determining current direction, make sure to ask in class.

We can also have many layers of cross beds (figure 5.5.16). In some regions, the cross beds within different layers can indicate different paleocurrent directions. This indicates to a geologist various information about the past changing environments. In figures 5.14 and 5.15, we are asking about what direction could have possibly formed in similar environments. However, figure 5.14 was taken at Desoto State Park in Alabama while figure 5.15 was taken at the Grand Canyon in Arizona. Clearly these two environments are currently very different, but now geologist can use clues like cross beds and some others that we will talk about to piece together a picture of what these past environments looked like.

In class, we talk about several other sedimentary structures. In this e-book, we are only going to touch on one more structure. This last structure will be fossil evidence. Fossils will provide a wealth of information about past environments and conditions based on the type of life forms that were living there. One easy example is the trilobite. We can think of these guys living in shallow seas in near shore environments. In the next course, *Earth and Life Through Time*, you

Figure 5.14 Can you tell the direction of the paleocurrent? Left or Right?

Figure 5.15 Can you tell the direction of the paleocurrent? Left or Right?

Figure 5.16 Many layers of rock beds.

will learn more about ancient life, but for now this assumption will work. In figure 5.17 we see a trilobite. This fossil is small; it is around the size of a penny. Any place you find this fossil, you can assume a marine environment. In figure 5.18, we are looking at lots of trace fossils. Trace fossils are fossils that indicate some life form has disturbed the sediments or has left behind some evidence of life.

Figure 5.17 Trilobite.

Figure 5.18 Trace fossils.

In figure 5.19, we see trace fossils of trilobites. We see the evidence of them moving across these sediments in the past. Look at the circle in figure 5.18. With the understanding that this is past evidence of a trilobite, we can now draw the assumption that this used to be a past marine environment. As you can see, it is currently not under water. This image was taken at the bottom of the Grand Canyon. This tells us that the sedimentary rocks found in the bottom of the canyon were deposited in a shallow marine environment. In figure 5.19, we can see a fossilized shell. This is not much different than the shells you will see at the beach during spring break. You are well aware what type of environment is needed to harbor life like this. This

Figure 5.19 Trace fossils of trilobites.

image was taken at the top of the Grand Canyon. I am sure you are aware that Arizona is not famous for its beaches. Again, it's another clue for geologists that help us understand the past environment.

Questions

1. What is a parent rock?

2. Describe saltation and traction.

3. What is the process we call sorting?

4. What are the categories and size distribution used to describe sediments?

5. What is Lithification?

6. What is stratification and how does it form?

7. What is a turbidite and how are they useful?

8. What are cross beds?

9. How can the asymmetry of a dune tell you past or present current direction?

10. How can fossils provide useful information about past environments?

CHAPTER 6

METAMORPHISM

THE CHANGING OF THE ROCKS

METAMORPHISM AND METAMORPHIC ROCKS

What is metamorphism?

Now that we have talked about igneous and sedimentary rocks, it is time to discuss how these rocks can be changed and end up with completely different properties. This change is called metamorphism. It means that the rock has been altered from its previous state, and is now referred to as a metamorphic rock. Both igneous and sedimentary rocks can be metamorphosed. In addition to these, rocks that were previously metamorphosed can be metamorphosed again. So the starting material of a particular metamorphic rock can be either an igneous rocks, a sedimentary rock, or a metamorphic rock.

Let's take a look at the simple example below that basically sums up metamorphism using everyday items. Here we see a simple mixed berry muffin mix. We are going to consider the dry mix to be a generic sedimentary rock.

In order to make these delicious muffins, we need to add milk. The milk will represent the fluids that are present in various sedimentary rocks or fluids from their aqueous environment. Now we take the muffin mix (hydrated sedimentary rocks) and place it in the oven. This act will increase the temperature of our mixture, and this temperature increase is what generates these tasty fluffy muffins in about 15-18 minutes. This step simulates the heat and pressure that rocks will experience from burial or exposure to hot magma/lava. The changes we will see in the rocks will take a lot longer than 15 min, but the result is the same. Look at the muffins, do they appear to have different properties now from when there were a liquid mixture? Sure they do! So when we add heat and pressure to rocks, we can expect our metamorphic rocks to also have different properties from our starting material. So in order to metamorphose a rock we need Heat and Pressure, Heat and Pressure, Heat and Pressure. One other factor that is often included

Figure 6.1 a-c "Morphing" dry mix into muffins: dry mix, add milk, bake to create baked muffins.

in this reaction is chemically active fluids. Think back to the milk we added to our muffin mix. When you heat the mix, the milk will evaporate, turn to steam, and play a role in the baking process. Like the muffin, metamorphic rocks will have a *change in texture and/or composition* as a result of the heat, pressure, and chemically active fluids.

Metamorphic rocks only cover ~15% of the Earth's surface and are generally found in mountain belts both old and new. The Appalachians are an ancient eroded core from a mountain chain that use to rival the Himalayas. So the Appalachians will be a great source for finding metamorphic rocks. Since the Appalachians extend into Alabama, our state has a fair amount of metamorphic rocks. A great example can be found along 280 extending from Auburn and decreasing in metamorphic intensity as one approaches Birmingham.

One particular example in our area are the rocks that are exposed at the base of the falls in Chewacla State Park (Figure 6.3). Many of you have visited this location, but little did you know that you were standing on metamorphic rocks that are well over a billion years old? The next time you visit, take a look at see if you can pick out any of the features we will be discussing in class.

Figure 6.2 **Appalachian mountains.**

Metamorphic rocks can be considered the result of alteration of any type of preexisting rocks. This alteration will occur within a specific realm of pressure and temperature. If the heat and pressure are too low, then the rock may not alter at all. If the heat and pressure become too high, then the rock will melt. If the metamorphic process melts the rock completely, then the resulting molten material will cool into an *igneous rock*. Once rock material is *completely* melted, it is no longer considered a metamorphic rock. So by having an upper and lower limit to metamorphism we can think of the process happening within a particular window of pressure and temperature. Note that I did say completely melted. . . ...we will talk about what happens if that rock is partially melted later in this chapter.

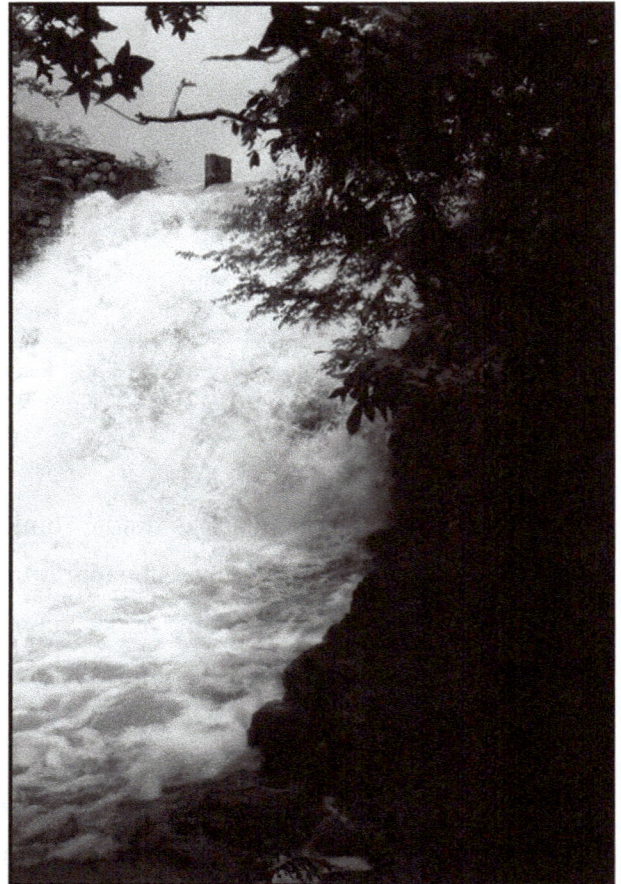

Figure 6.3 **Chewacla State PPark water fall.**

In order to generate a metamorphic rock, we mentioned earlier that we need primarily heat and pressure. Where does that heat and pressure come from? I am sure most of you have already guessed correctly that the heat and pressure will come from within the Earth. Due to the heat engine at the center of the Earth, the deeper you go within the Earth, the hotter it will become. This is also known as a geothermal gradient. The geothermal gradient relates temperature to depth. In general, we consider this to be 25 degrees centigrade per kilometer. Hence the deeper you go, the hotter it gets. This is a general average and will vary depending on if you are working with continental or oceanic crust. This idea will allow us to make predictions about the maximum depth of a rock based on its metamorphic grade.

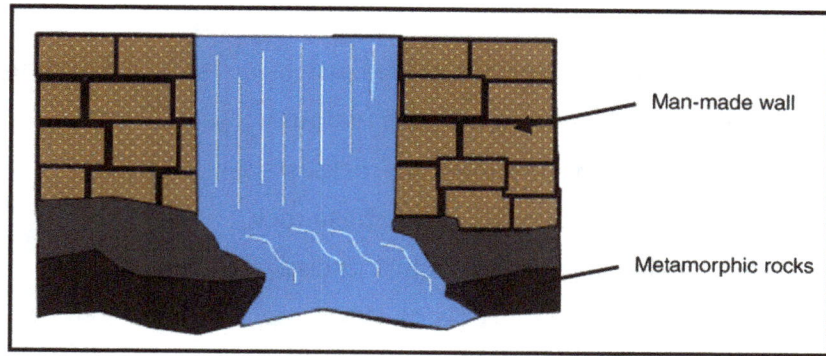

Figure 6.4 **Metamorphic rocks from man-made wall.**

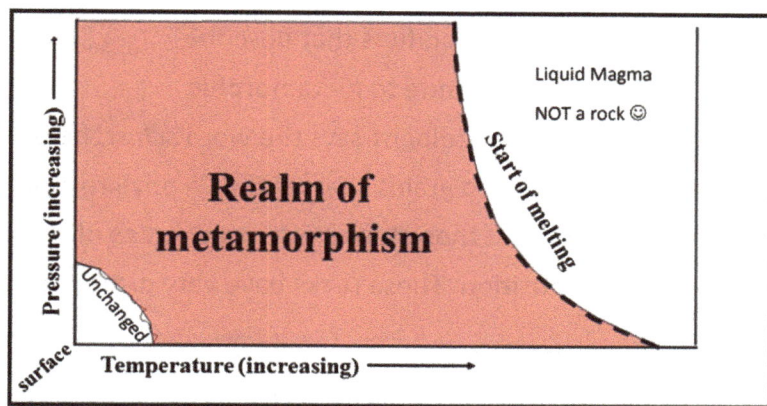

Figure 6.5 **Realm of metamorphism**

What is metamorphic grade?

Metamorphic grade is easy! Imagine you are studying for your metamorphic test, and you put in lot of time and energy studying. Congratulations, you made a high grade on the test. In contrast to your study buddy who only put in small amounts of time and energy, they made a low grade. Just like you, the more temperature and pressure (time and energy) a rock experiences the higher its metamorphic grade. So we can then assume that the deeper we go we will find rocks of higher and higher grade until we are deep enough for the rock to melt completely. In metamorphosed sedimentary rocks, this will produce rock types of various grades just as this class will have students with various grades (A-B-C-D-F). In sedimentary rocks, starting with unmetamorphosed shale, these rock types are; shale-slate-phyllite-schist-gneiss (highest grade of metamorphism). What separates these four rock types is their texture. As metamorphic grade increases we will see an increase in grain size. Please keep in mind that these are textural modifiers that describe the increase in grain size due to metamorphic

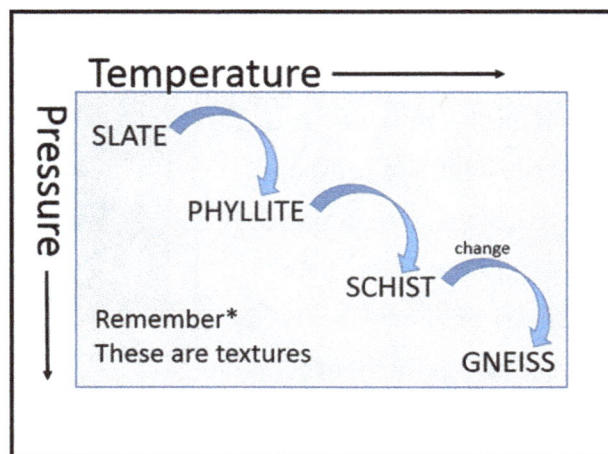

Figure 6.6 The more temperature and pressure a rock experiences, the higher its metamorphic grade.

conditions. So when a geologist says the word *schist,* he or she is referring to a metamorphic grade that has produced grains coarser than a phyllite, but lacking the texture of a gneiss. It is important to note that these terms refer to degrees of metamorphism, and are not bound by any specific composition. These rocks have very distinct appearances, and you will work with all types in lab.

What is metamorphic foliation?

Metamorphic foliation will be present in the majority of metamorphic rocks. We will talk about the other small percentages later. Metamorphic foliation will appear as lines on the surface of the rock. It is important to remember that in the world of geology we have to think in 3D, and these lines are not actually lines but planes. Think about it this way; imagine you are looking at a stack of plates, from the side it looks like lines in the surface of the rocks. You are familiar with a stack of plates. You know the plate extends through the stack and not just expressed as a line on the surface. The same goes for metamorphic foliation.

So how can metamorphism generate these planes? Imagine this scenario, its summer and you are out riding in the country with the windows down. You stick your arm out the window to feel the wind rush through your fingers. If you hold your hand perpendicular to the flow of wind, you will fell the wind resistance pushing back on your hand. If you rearrange your hand into a different orientation, say parallel to the wind, you can avoid this wind resistance almost

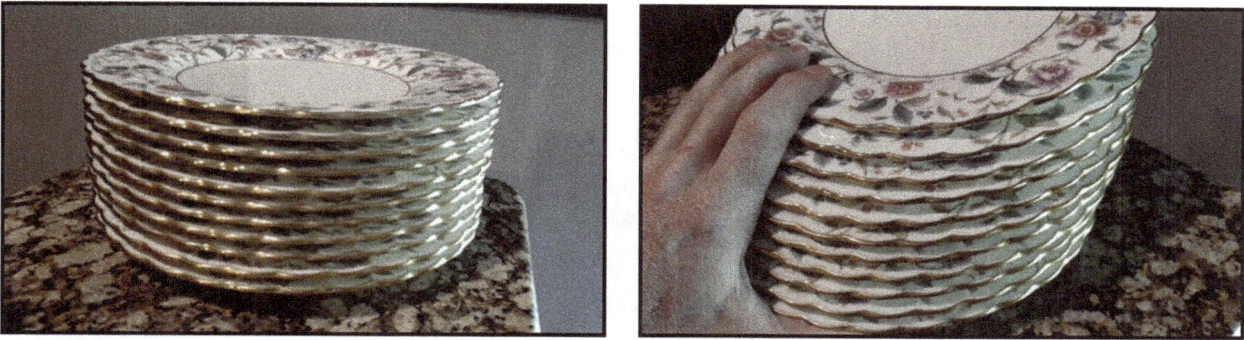

Figure 6.7 a-b Stack of plates illustrates that lines on rock surface are actually planes.

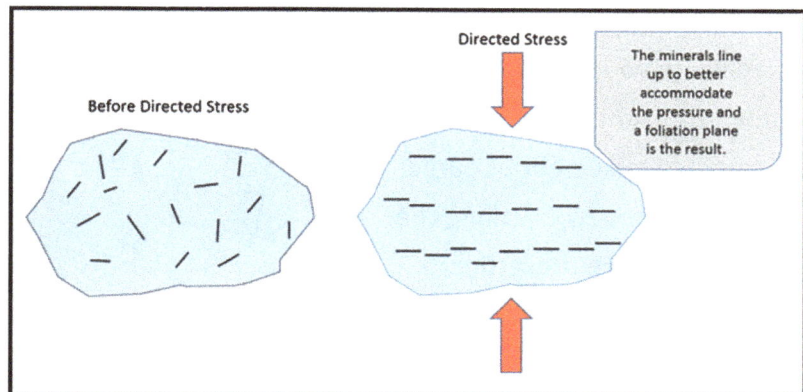

all together. Minerals do the same thing. When a mineral is under directed stress, you can think of it as if they are experiencing the wind resistance mentioned above. They will change their orientations (just like you turning your hand) to better fit into the stress field the find themselves in. Since

Figure 6.8 Minerals change orientation with subjected to directed stress.

all the minerals in the sample are experiencing the same pressure field, they will all line up in roughly the same way. This lining up is how metamorphism generates foliation.

How do Rocks metamorphose?

So we know that we need heat and pressure, we know this heat and pressure can and will change our texture, we know how this change will produce foliation, and we know that any pre-existing rock can undergo metamorphism. So what actually happens during the metamorphic process? We have two main categories in which metamorphism can change our rocks. It can do so during *mechanical dislocations* and/or *recrystallization*. During mechanical dislocations, the crystal lattice is actually broken or deformed. This results in the rock breaking or smearing while it is experiencing metamorphic forces. The best way to understand this process is by thinking about waffles. This is the same analogy that I use in class, but I think it works very well. So let's talk about waffles.

I want you to imagine that you run to your kitchen, open your freezer, and grab that box of frozen waffles. If (for some crazy reason) you wanted to break that waffle in half, you could do so. The little squares that you fill with syrupy goodness will break right down the middle. Now imagine that the grid pattern you see on that waffle represents the crystal lattice that we talked about earlier in the semester. By breaking the waffle, you broke the lattice; you broke (mechanically dislocated) your rock! Would you have been able to break this waffle if it had not been frozen? Probably not. Mechanical dislocations that behave in this manner are referred to as *brittle* dislocations. Breaking glass, hard candy, even your frozen waffle all display brittle dislocations. The unifying factor—they are all at lower temperatures. Temperature is the key as to how the mechanical dislocation will behave. Once we begin to increase temperature there will be a noticeable change.

Figure 6.9 a-c When breaking a frozen waffle, the grid pattern gets broken (mechanically dislocated).

Figure 6.10 In this microscopic view of the internal structure of a deformed rock, we can see an example of ductile deformation. If you look at the "smeared" mineral that is just above the plagioclase (mineral that has the stripes) you can see how that grain was smeared and stretched. Indicating this was at higher temperatures when this deformation occurred.

Let's head back to our waffle example. Same scenario as before, you run into your kitchen and open to freezer only to find that your roommate left the box of waffles out overnight. You are devastated, but you still reach in and pull out a soggy waffle. This waffle will be limp in your hands. If you wanted to brake it in half, the little squares would only smear rather than break. This smearing behavior is what we call *ductile* dislocations. The warmer the rock becomes, the more likely that the material will behave more ductile rather than brittle. So this helps geologists by providing context clues to the history of a rock. For example, if a rock has ductile deformation then that tells us the rock must have been hot and much deeper in the Earth. Whereas brittle deformation in a rock would tell us just the opposite.

In addition to dislocations, metamorphic conditions can also cause the rock to recrystallize. Two types of recrystallization can occur. The ***first type*** is an easy one. This is just where a type of grain will recrystallize into a new grain of the same material. So what does that mean? For example, if you have a bunch of loose quartz grains, and let's say they

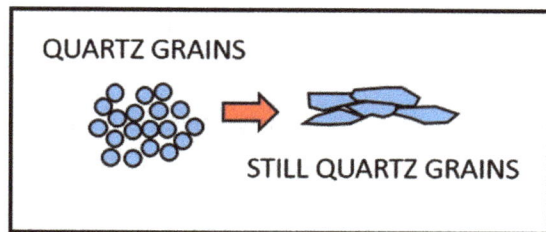

Figure 6.11 **Grains can recrystalize**

are all sand sized and sub-rounded. If these grains were to be recrystallized, they would still be quartz (the sample lacks the raw material to grow into anything else), but they would change and essentially grow together (recrystallized) in favor of the metamorphic stresses that the material is under. This type of recrystallization is what will change a common sand stone into a quartzite. You will work with both of these samples in lab, and I want you to think about the recrystallization process when working with these samples.

The ***second type*** of recrystallization is a bit more exciting. In this situation we will actually be growing new minerals. We will be using the raw material that is available within the preexisting rock to recrystallize it into totally new minerals that can only be formed during metamorphism. We will call these our metamorphic index minerals. In class, and for this example, we are going to be working with 7 metamorphic index minerals. Now, one of these 7 will actually pull double duty. It is a mineral that can form by metamorphic and igneous processes. So when you see it listed below, just remember that it is a busy body and just like to be in everyone's business. *The process of mineral growth due to recrystallization is a complicated one.* In this course we are only going to look at one sequence that is intended to help us grasp the concept. Just be aware that we are only going to focus on minerals that will grow when we metamorphose shale. Luckily for us, there is a lot of shale in the world.

As I mentioned earlier, there are 7 minerals that we are concerned with in this process. I am going to list them and then we will talk about why they are important. I think it would be easier if we break this list of 7 down into two lists, one of 4 and one of 3. You will see why in a few moments. We will talk about our list of four minerals first: chlorite (a green sheet silicate, it kind of looks like green muscovite), biotite (this one you know, you had it in lab and this one is pulling double duty), garnet (we had this on in lab, and this is also your favorite mineral), and staurolite (we also worked with this in lab early on, this one made the little brown crosses).

Now if we look at Figure 6.12, we will notice that the controlling factor in this sequence is temperature. You can clearly see that chlorite will form at a lower temperate than garnet. So if you are a geologist and you find a rock that has biotite in it, you will be able to make the assumption that the rock was at a temperature high enough to allow for this type of

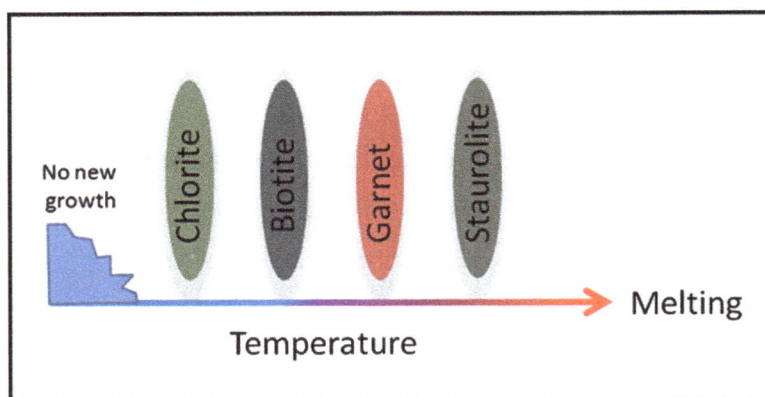

Figure 6.12 **Temperature affects crystallization.**

recrystallization. At this point, this is clearly just a qualitative statement. We will soon add actual temperature requirements that will allow us to make more precise observations about temperate and the growth of these minerals. As of now, if you understand that the temperature must increase to grow the minerals to the right on the chart, then you are right where you need to be. Now let's look at the other 3 minerals that we will need.

The next sequence of minerals is kyanite, sillimanite, and andalusite. You have worked with kyanite in lab and the other two are what we call polymorphs of kyanite. A polymorph is a mineral that has the exact same chemical composition, but due to either temperature or pressure, the structure is altered. If you remember back to the 5 rules for minerals, you will recall that if we have a different structure, then we will have a different mineral. So these three minerals all have the same chemical formula (Al_2SiO_5) but are polymorphs of each other. In this case we can use the variety of polymorph to help us determine the amount of pressure the mineral was under. In Figure 6.13, you can see that we now have a pressure bar on the left hand side. This is defining pressure as increasing upwards toward the top of the graph. Let's take a look at an example. You have a rock sitting at 750 degrees Celsius at ~2kpars of pressure. What mineral do you think we would find at this location? Looking at the graph, we can plot that point and tell that we are going to find andalusite. Now what would happen if I began to increase the pressure and had the temperate remain constant? If I increased it to 5 kbars, I would now have sillimanite. That previous mineral that was andalusite would react to the increasing pressure, and it would alter its chemical structure. So what was andalusite at 2 kbars will be sillimanite at 5 kbars. If we continued to increase the pres-

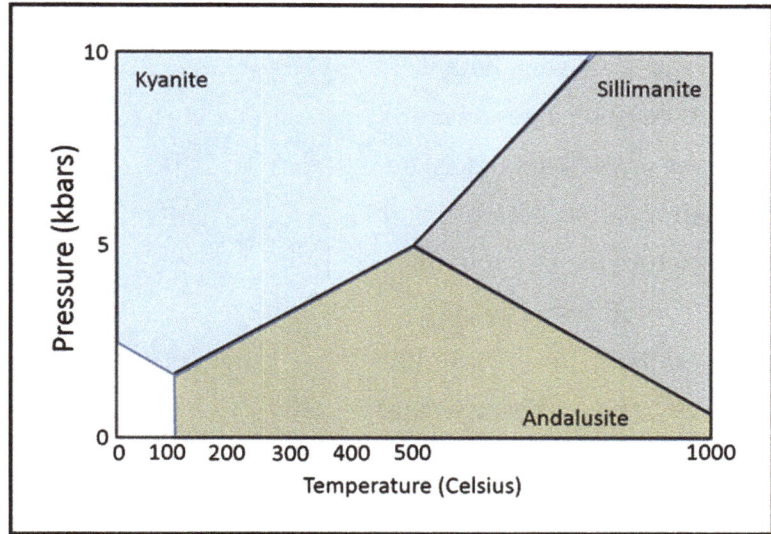

Figure 6.13 Kyanite polymorphs due to pressure and or temperature.

sure up to near 10 kbars, this same material would alter its chemical structure again and form kyanite. So you can see that the main controlling factor for these three minerals will be pressure. Temperature does play a roll, but we can use these minerals as a tool to help use make assumptions about its maximum pressure.

So to understand the two sequences of mineral recrystalli-zation, we can use one for tem-perature and one for pressure. In order to do that we must now combine the two graphs.

So now we are set to make assumptions about our meta-morphic rocks by using this new tool. For example, if I handed you a rock with garnet and kyanite present in the sample. You can now look at the chart and tell me

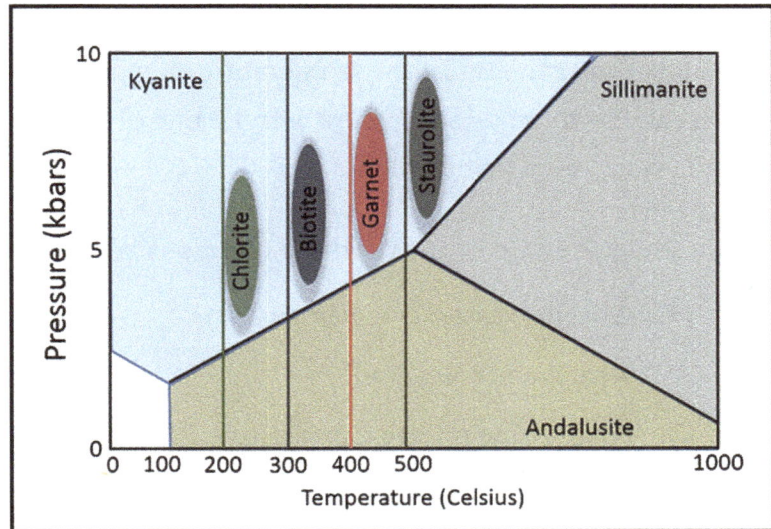

Figure 6.14 Both graphs combined

that in order for garnet to grow, the sample had to be at least 400 degrees Celsius and then you could tell me that in order to have kyanite existing with it, you would need a minimum pressure of ~4 kbars. If I asked for the minimum pressure and temperature for a rock with chlorite, biotite and kyanite, you would answer at least 300 degrees C and ~3 kbars. In that previous example you want to give me the minimum temper- ate for the highest grade min-

Figure 6.15 What minimum temperature and pressure are required to produce this sample?

eral. Biotite comes after chlorite, so the chlorite becomes less important. You want to focus in on the highest grade mineral. Now for one last example, see Figure 6.15 and see if you can give me a minimum temperature and pressure.

The minerals that are obvious are biotite and garnet. So that should make guessing the temperature rather easy. What about the pressure? Well if I do not give you one of the polymorphs, you do not have the information needed to make that assumption. If no polymorph is listed (for this course) you can just list the temperature. In more advanced studies there are tricks that geologist can use to arrive at pressure without one of the polymorphs, but for this class we are not going to worry over that detail.

You should be able to answer these questions at this point.

1. What is required for metamorphism?

2. What is metamorphism?

3. What are the two types of mechanical dislocations?

4. What are the two types of recrystallization?

5. What are the 7 index minerals that we use for determining pressure and temperature?

6. What is a polymorph?

7. What is the minimum pressure and temperature for a rock with staurolite and kyanite?

We have been talking about metamorphism in the sterile environment of our classroom. If you are a geology major, or if you decide to become one, you will be have to deal with these minerals in their natural setting. Often in class we will pick the very best examples that we can for teaching purposes. This is great because it helps you learn, but when you get in the field you might find that not everything

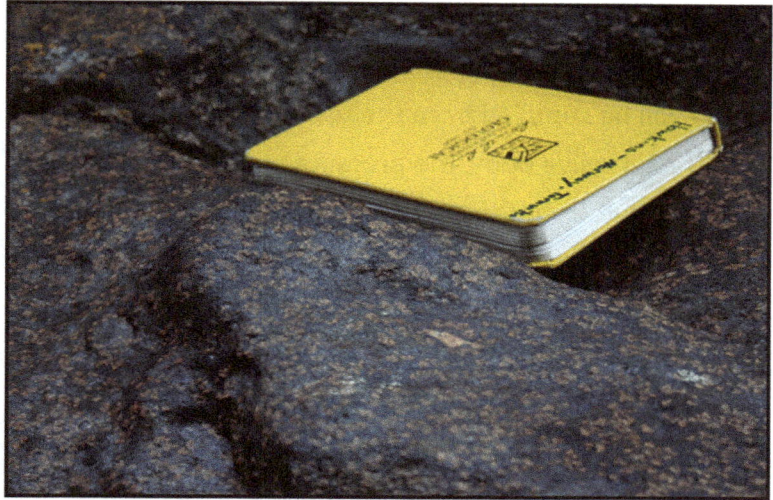

Figure 6.16 Some rocks are difficult to identify.

looks as crisp and clear as it does in class. Figure 6.16 was taken while I was conducting field work in Norway. It is still better than the examples you might find around here, but it gives you an idea of the point that I am trying to make. This rock is also garnet and biotite, but as you can see it might not be as easily identified.

Types of metamorphism

Now that we have learned about the processes involved in metamorphism, we can now apply those to the three broad types of metamorphic events. These types are: contact / thermal, regional / dynamo-thermal, and dynamic. These three types are very common but the amount of material affected by each varies. Regional metamorphism is the most abundant and accounts for the highest volume of metamorphic rocks. The rocks here at Auburn University are the result of regional metamorphism. The second according the volume is thermal. We have some examples that are semi-local, but this type is limited in scope. The third type, dynamic, is by no means less important. We do have some local examples, however, it only affects as small amount of material. We will start this section by looking at contact metamorphism first.

Contact metamorphism

Contact metamorphism is exactly what the name implies. This is the metamorphism of material that is directly due to coming in contact with a heat source. This is generally achieved when there is some form of igneous intrusion (pluton, dike, or sill). In this process the heat required will be supplied from the residual heat from the intrusive body, the pressure will be a direct result from the depth of this intrusive body. In this process heat is the primary player. Yes, pressure and chemically active fluids are present, but it is the introduction of this heat source that starts the process. Over time, the heat within the intrusive body will penetrate the cold county rock and create an area called the *Contact Aureole*. This contact aureole (CA) is the region immediately adjacent to the heat source where the metamorphism is actually occurring. You can see in Figure 6.17 that this is a small region when compared to the size of the intrusion. This is why contact metamorphism only accounts for a small volume of rock. Within this CA you will find that the minerals will grow in zones based on the temperatures that we discussed earlier in the chapter. For example, the section of the CA that is in contact with the intrusive body will be the hottest. Here we could say that the settings are correct to form staurolite. Now as you move away from the heat source, the temperature will drop. So there will actually be rings of minerals forming due to the lowering of metamorphic temperature.

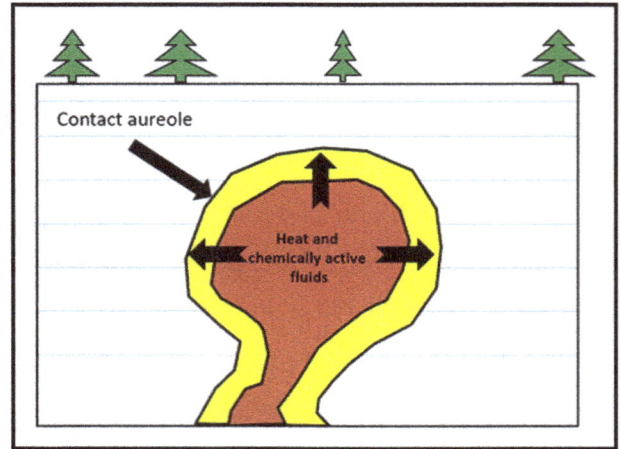

Figure 6.17 Heat penetrates cold country rock to create an area called contact aureole.

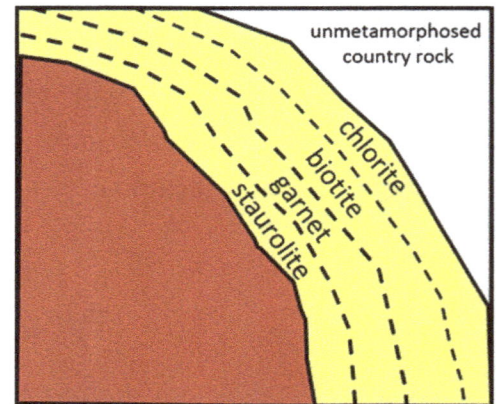

Figure 6.18 The contact aureole is where the metamorphism actually occurs.

There is one other important observation about contact metamorphism that we need to talk about before moving on. Contact metamorphism will not produce foliation in the affected rocks. If you recall, foliation forms due to directed stresses that cause the mineral grains to line up. In this case, we are only dealing with stress from the rocks above. This will not generate any metamorphic foliation. In lab you will have several samples of metamorphic rocks that do not have foliation. I want you to take a look at these and think about how they might have formed. Generally, foliation will be produced by the next type of metamorphism.

Regional metamorphism

Regional metamorphism or sometimes referred to as dynamo-thermal metamorphism will affect rocks over large regions. This is the type of metamorphism that can be found acting along the edge of a continent that might stretch for 1000s of miles. So you can see that this type will far exceeded the volume affected during contact metamorphism. This type of metamorphism is found in mountain belts. This is the process due to plate tectonics and the massive amounts of pressure and temperature that is generated when continents collide. We will find metamorphic rocks that are well foliated and extend for 100's of miles from the point of highest metamorphic energy. In class we will talk more about this particular metamorphic event.

Dynamic metamorphism

Dynamic metamorphism is the result of rapid movement within the rocks. This is generally found as the result of meteorite impacts or within fault zones where rocks are sliding past each other. In Wetumpka, Alabama we have an ancient impact feature that was created ~80 million years ago. Even though it occurred in the distant past, geologist have found rocks from deep within the heart of the impact feature that show the affects of the enormously high pressures and temperatures generated during this brief event. These events are not common in modern times, but the Earth does have many scars from past impacts. All of these will have dynamical metamorphosed material.

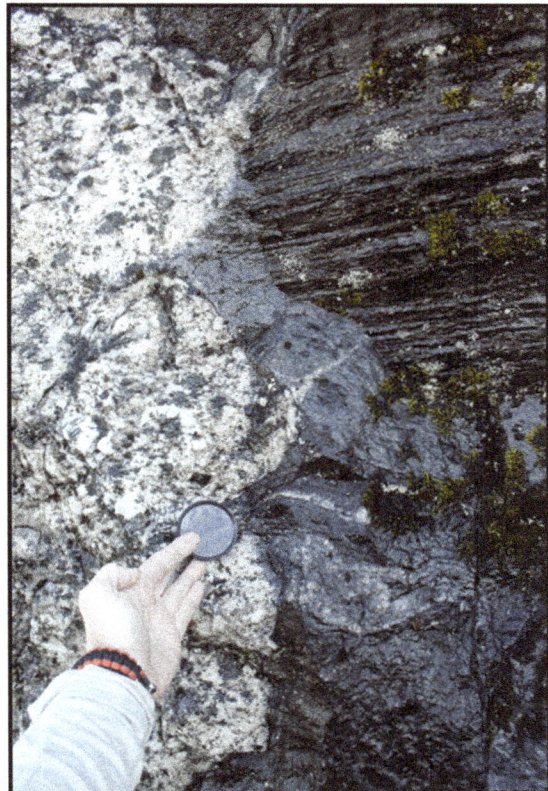

Figure 6.19 Regional metamorphism found in mountain belts.

The more common producer of dynamic metamorphism is fault movement. Fault movement will generally only affect the rocks within the fault system, so it is not a large volume of material. Luckily for geologist, there are lots of faults. We will cover faults in more detail in later chapters, but for now we are going to define faults as any situation where rocks are sliding past each other. It is this motion that will generate the mechanical dislocations that are associated with this type of metamorphism. We can have two types of mechanical dislocations caused by faults, and the controlling factor will be temperature. This goes back to our waffle analogy. If the rocks are cold, it will shatter and destroy the crystal lattice when the rocks move. We call this a cataclasite. In Figure 6.20, we are looking at a thin slice of a rock that was affected by dynamic movement while it was cold. This resulted in a cataclasite and you can see in the image where the individual crystals were shattered. This is considered to be a brittle type of movement.

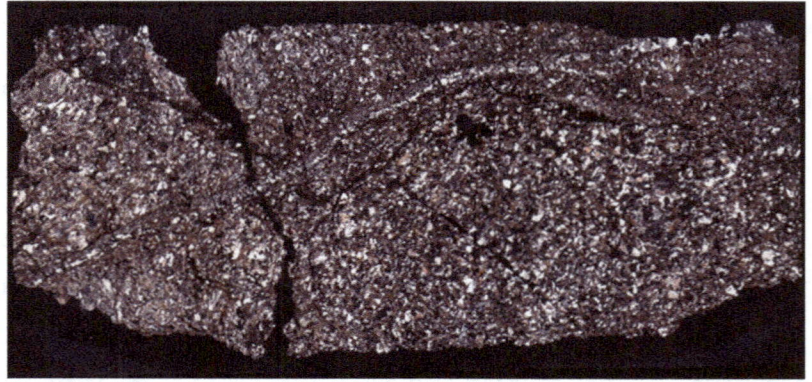

Figure 6.20 This sample as affected by dynamic movement while it was cold resulting in a cataclasite.

The other example, warm or ductile movement will produce an entirely different looking material. This material will appear to be smeared and rolled. We call this type of dynamic metamorphism a mylonite. In Figure 6.21 of the

Figure 6.21 Mylonite is formed by dynamic metamorphism from warm or ductile movement.

mylonite you can clearly see how the larger pink minerals have been smeared and stretched. It is a stark contrast to the shattered crystals found in the above example. Geologist can use this difference in texture to make assumptions about the depth of the rocks when this deformation occurred. Deeper rocks will form mylonites, while rocks near the surface will form cataclasites. We will use a general rule of thumb of about 15km as our cut off. So if you see ductile deformation on our next test, you can think deep warm rocks that were at least 15km deep.

Extreme metamorphism

We have talked about how rocks change when placed under escalating pressures and temperatures. But what happens if the temperature and pressure gets too high? The rocks will start to melt or at least it will allow for the migration of ions. In class we are going to talk about three very important vocabulary words. These are: gneiss, anatexis, and migmatite. Gneiss is easily spotted because it will usually have light and dark bands that were generated because the rock got hot enough for ions to begin migrating. When ions of similar type join together, you will see these bands forming. If the temperature and pressure continues to rise, we call it anatexis, or extreme metamorphism. This is where part of the rock will actually melt. We will talk in class about why some minerals will melt while others remain solid (think back to Bowens RXN series), but the resulting melted material will solidify into an igneous rock. So we can actually get it hot enough to melt less stable minerals within a rock. This melting out of material is referred to as anatexis. The resulting rock will have the remaining stable metamorphosed material as well as the newly solidified igneous material. When a rock has both igneous and metamorphic material, and that igneous material is derived by extreme metamorphic conditions, we call it a migmatite. The enclosed picture is a migmatite. Can you identify the igneous and metamorphic material?

CHAPTER 7

PLATE TECTONICS

THE BUILDING OF MOUNTAINS

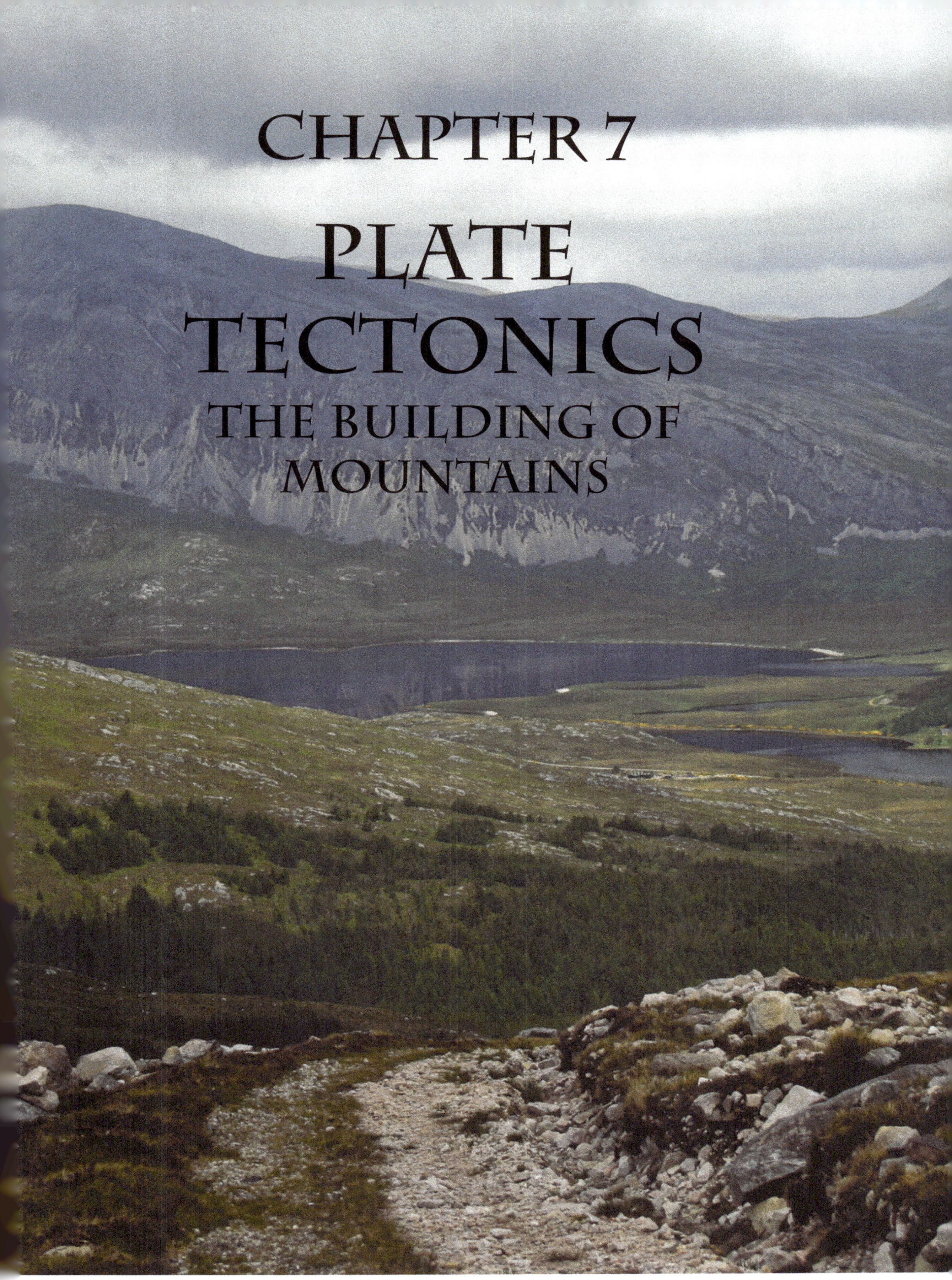

When you are confronted with the idea that our continents are moving you probably do not produce much opposition to the idea. This is now such an accepted fact that making sense of our world without plate tectonics would even be more difficult. This; however, has not always been the case. Most likely your grandparents grew up in a work that did not believe in the idea of plate tectonics. You can easily accept that our continents are moving a few centimeters a year, but for the people living pre 1960s, this would have been absurd. How can something as large as a continent move? It was that very question that held back the acceptance and understanding of plate tectonics for decades and personally shamed Alfred Wegener.

Alfred Wegener was a scientist that was primarily interested in meteorology, and he was a pioneer of polar research. He was born in 1880 and died in the artic in 1930, and he is still buried in an icy tomb. During his meteorological research he started to notice information that he believed suggested that the continents were once all connected. This idea did not come to him over night, but slowly formed after years of evaluating his 5 pieces of evidence.

In class we will talk about each one of Alfred's observations, and how they work geologically. We will look at them critically just as the other scientists of his time would have. Alfred Wegner published his idea that he based on these five observations in 1920. He called his idea "The Origins of Continents and Oceans", but this has been distilled down to the idea of Continental Drift. The evidence is sound, and at first it seems like he makes a good argument. However, the question of what mechanism could there be to actually move a continent was a constant hurdle for his claim. The fact that no such mechanism could be described caused his idea to fall out of favor and eventually be forgotten. It was not reenergized until the late 1950's with the advent of new technology used in exploring our ocean floor.

Wagener's Evidence

1. The shape of the continents seem to fit together.

2. The structures and rock types match all the edges of continents that fit together.

3. Locations of past mid latitude coal swaps suggest different continental locations in the past.

4. Glacial striations point to a different continental configuration in the past

5. Fossils seem to have geographic ranges that extend past what would be natural barriers.

After WWII the US government decided it would be prudent to study and map the bottom of all the oceans. It was assumed at the time that the ocean floor was a flat and featureless place that would have little if any topographic variation. With the introduction of SONAR, it became possible to produce an accurate topographical map and see just how featureless the oceans were. Turns out, they were not featureless at all. Instead they found mountain ranges on the bottom of the ocean. These mountain ranges were 1-3 kilometers high and created an inter-connected chain that ran for more than 115,000 kilometers. We call this unbroken mountain chain "mid-ocean ridges". All the oceans have them, and they are formed by a very important function.

In addition to mapping the topography, they wanted to map the remnant magnetic field of the rocks on the ocean floor. Why? Submarines use detailed magnets to navigate the oceans, and if the rocks could be throwing them off, it would be important to know. So the why is easy to understand, but how do rocks affect magnets? Turns out that since mafic rocks have high iron content, that iron can preserve magnetic fields. You know that rocks can be magnetic, you had magnetite in lab this semester. The rocks on the ocean floor are on a much weaker scale with respect to

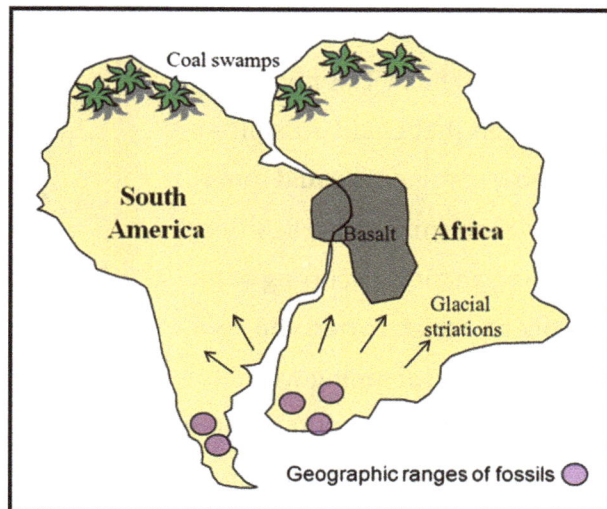

Figure 7.1 "The Origins of Continents and Oceans."

the magnetic force. When rocks are molten the little iron rich minerals will react to the magnetic field of the planet, much in the same was as a compass would. These little minerals line up, and when the rock hardens, it has preserved a record of the magnetic field of the Earth when the rock was molten. Okay, two problems with that. One, why is there molten material on the sea floor? Two, they found that the remnant magnetic field preserved in the rocks was not consistent. They found that it occurred in stripes.

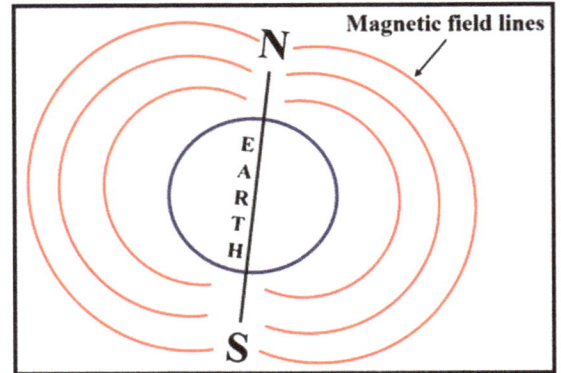

Figure 7.2 Earth's magnetic fields.

The stripes reviled two pieces of information. They reveal that in order for these rocks to be recoding magnetic fields of the present and the past, that there must be a constant flow of new molten material being added at the mid-ocean ridges. So now we can see that Earth's surface is gaining material and crustal amounts are growing at mid-ocean ridges. The other major reveal that we are not going to dive into in this course is that Earth's magnetic field tends to switch polarity for long periods of time in our geologic past. We are not sure why, and we are not sure how. We just know that it happens. Currently our compasses point north, if we were to experience a polarity reversal then they would point south. The take away message is that new material is being added to our ocean floor. This became a mechanism that could possibly explain how continents could move. This mechanism is called sea-floor spreading and it is occurring at all mid-ocean ridges. When you combine sea-floor spreading with the theory of continental drift, the theory of plate tectonics is born.

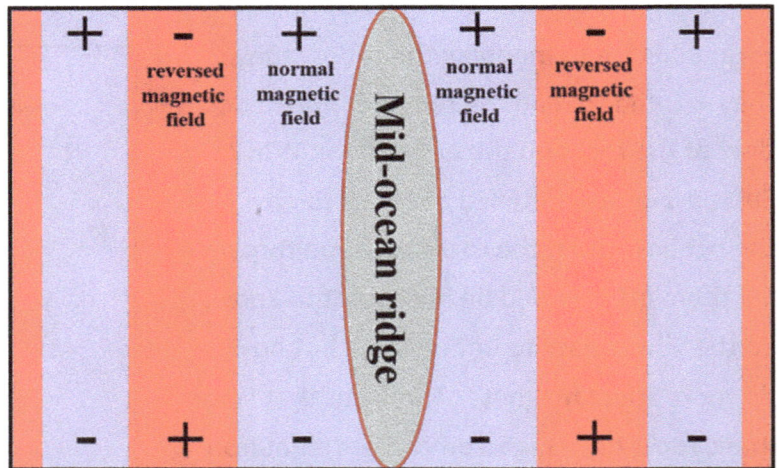

Figure 7.3 Magnetic field of the ocean.

Plate Tectonics

The idea that forms the basis for plate tectonics (PT) is that the Earth's crust is broken into many plates. These plates are mobile. They move due to convection cells working deep within the mantel that gently push on these plates and move them around the globe. Some of these plates are massive and contain our continents. For example all of North America is contained on the North American plate. Some are tiny, no more than a few 1000 kilometers in size. In class we will look at these different plates and how their movement is affecting our continents. I know that does not sound small, but compared to all of North America, it is. Plate tectonics is a very detailed and complicated system with many moving parts. In this course we are going to break this idea down and focus on plate to plate interactions. There are only three ways that these plate boundaries can interact. One, they can come together. We call this plate interaction a convergent boundary. Any place where two or more plates are coming together there will be compressional or convergent characteristics. Two, they can move apart. This boundary is called a divergent boundary. This will stretch and pull the plates apart. In this example in the void left from the plates moving away from each other, new material reaches the surface. This is like the mid-ocean ridge we described earlier. Three, they can slide past each other. We call this type of boundary a Transform boundary. These are not as common at the previous two but the ones that we do have are very important. Where plates slide past each other is generally a hot zone for earth quakes. We will look at these three boundary types in detail below.

Divergent Boundaries

Divergent boundaries are caused by tensile (pulling apart) stresses from beneath the tectonic plate caused by the movement of convection cells. This process will cause the plate to stretch and thin, potentially dividing the plate into two. This can occur below continental or oceanic crustal types. If this processes initiates beneath continental crust, then the process can result in the formation of a new ocean due to thinning of the continental crust, and if the process is beneath an ocean currently the sea floor will be spreading. In Figure 7.4, we can see a divergent boundary at letter A. You can see that the forces from the convection

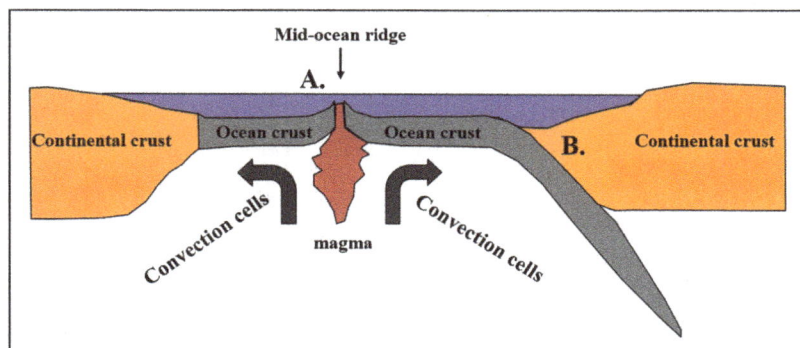

Figure 7.4 **This graphic highlights the convection cell movement with the two main types of boundaries. You can see both the divergent and convergent boundaries in this diagram. The Earth is not growing in size, so we know that the rate of subduction matches that of sea-floor spreading.**

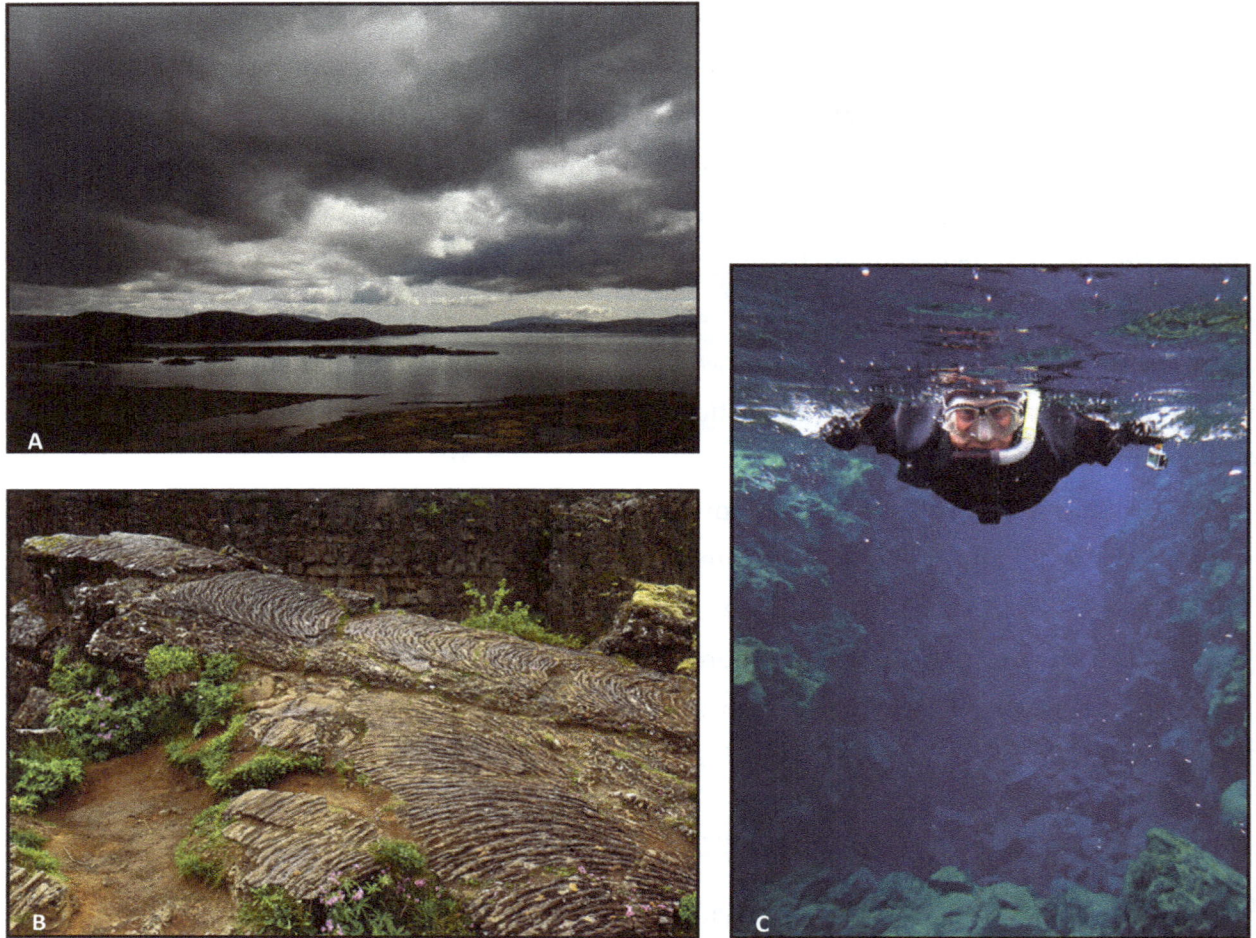

Figure 7.5 a-c The valley of Pingvellir in southwest Iceland. A: Valley with mountains in background. B: Recently hardened lava. C: The rift at work: between North American and Eurasian plates.

cells are pushing the plates apart allowing for the new molten material to make its way to the ocean floor. This is how we generate the molten material that records the magnetic stripes we talked about earlier. The best place to observe an oceanic crust divergent system is on the island of Iceland. In this special location we can actually see the divergence in action. Here the molten material has generated a landmass that extends above sea level.

One of the best examples of this can be seen in the valley of Pingvellir in south west Iceland. In photograph 7.5 A, you can see the valley with its mountains in the background. Due to the tensile forces at play here, the valley has dropped down leaving the mountains behind to form the eastern wall of the valley. In photograph 7.5 B, you can see some recently hardened lava material that was flowing within this divergent center. If this had been submerged, this would have been new sea floor instead of a hiking path. In photograph 7.5 C, you can see the rift at

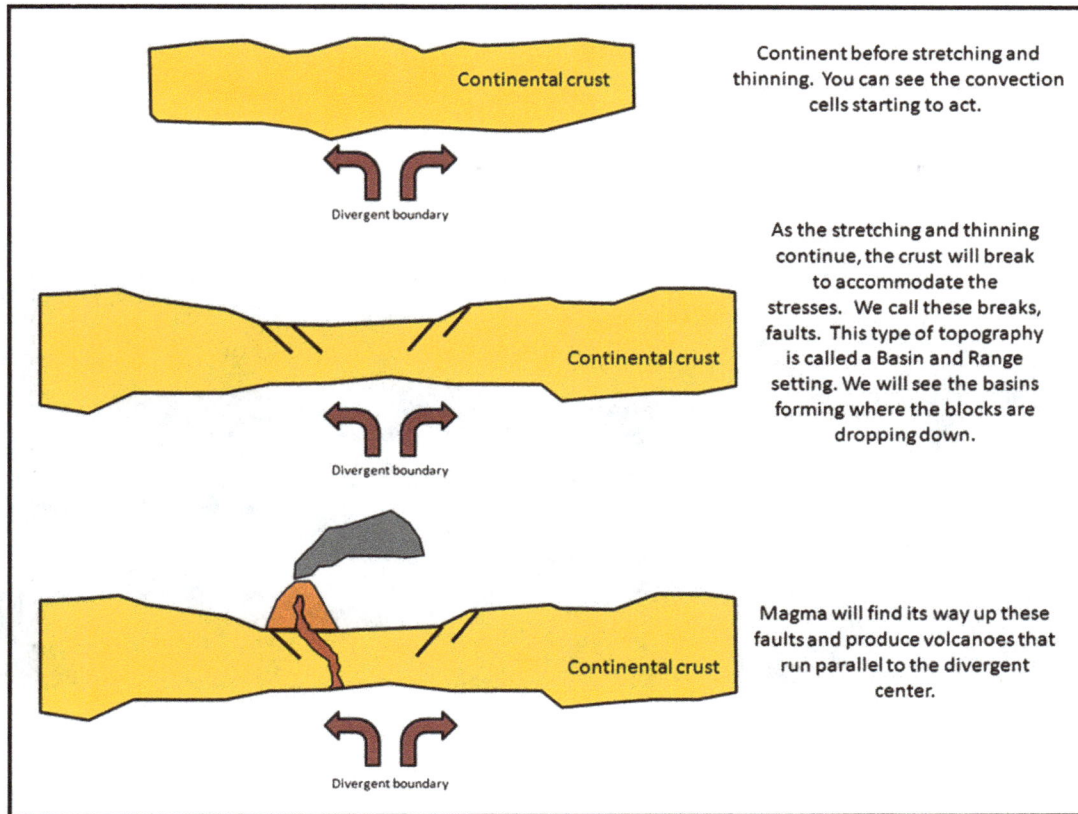

Continent before stretching and thinning. You can see the convection cells starting to act.

As the stretching and thinning continue, the crust will break to accommodate the stresses. We call these breaks, faults. This type of topography is called a Basin and Range setting. We will see the basins forming where the blocks are dropping down.

Magma will find its way up these faults and produce volcanoes that run parallel to the divergent center.

Figure 7.6 **Steps in the division of a continent.**

work. To my right hand side are rocks that belong to the North American plate, and to my left are rocks that belong to the Eurasian plate. So you can clearly see that divergent boundaries have the power to divide these massive plates. In Iceland we see a mid-ocean ridge system exposed. What does it look like if this divergent/rifting center develops beneath continental crust? In figure 7.6 we are going to look at the steps in the division of a continent. I want you to think about any similarities between the divergent boundaries seen in Iceland with those acting below a continent, like Africa for example.

This is currently the setting that you will find in east Africa. It is called the East African rift system. The rift currently begins in the country of Ethiopia and extends southward down into Tanzania. In Tanzania we can see the down dropped blocks just like we see in the Basin and Range province of the southwestern US. In Figure 7.7 we can see a valley as you are looking off towards the east. This is the Gregory rift valley in central Tanzania. This valley has dropped down just like the valley we see in Pingvellir. Along the scarp or valley wall, we would find a fault. This valley is what we would call a basin and the high elevation where I took the photograph would be called

Figure 7.7 Gregory rift valley in central Tanzania.

Figure 7.8 The Ngorongoro Crater.

the range. In addition to drop blocks, Tanzania also has its fair share of volcanoes. There are active and extinct volcanoes associated with this rift system. Volcanoes in this region can grow to be very large. The entire Ngorongoro Crater is the remains of one of these ancient volcanoes.

Convergent Boundaries

Convergent boundaries are plate interactions where two plates are brought together. This will generate compressional forces, so it is just the opposite of the extensional forces we observed in convergent boundaries. Whereas before we have crustal thinning, now we will see crustal

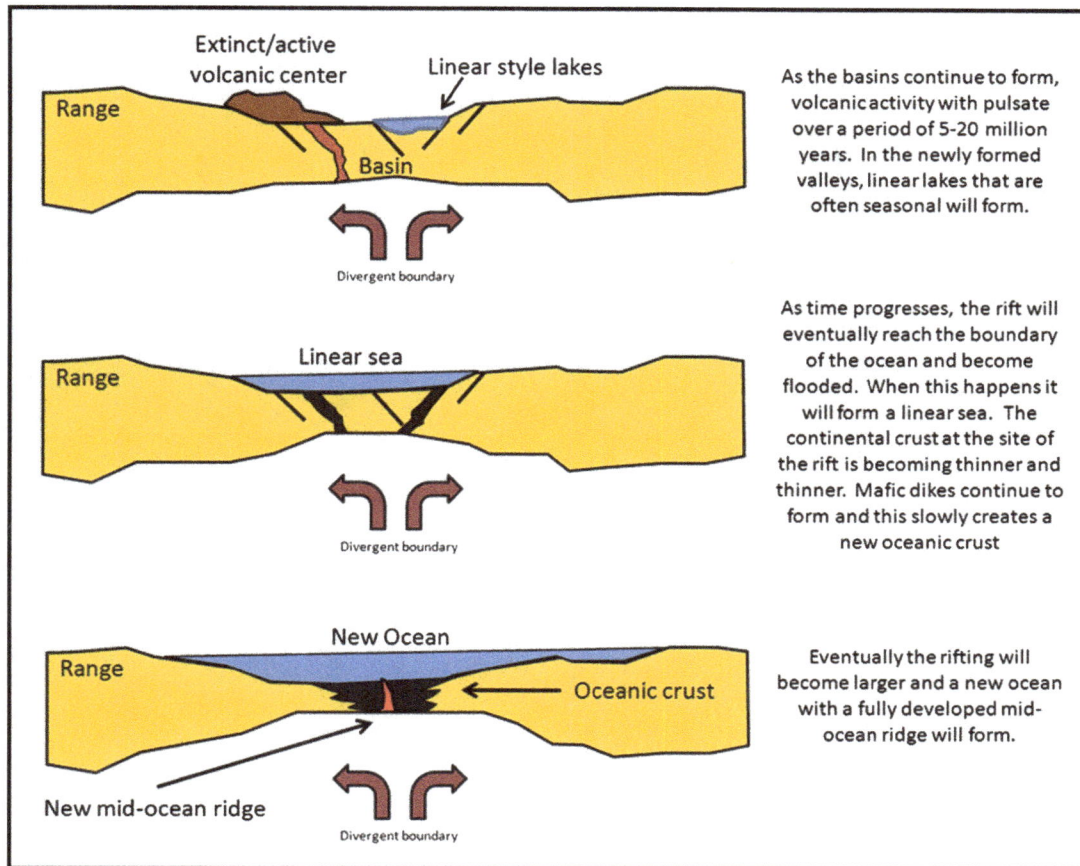

Figure 7.9 Additional steps in the division of a continent.

shortening/thickening. This is generally referred to as the mountain building process. Many of the mountain ranges around the world have been generated by convergent boundaries. You can imagine when two very large land masses collide that there would be carnage, and our closest example would be the Appalachians. Remember that we mentioned them before in our metamorphic chapter as regional metamorphism. Convergent boundaries are our primary means by which to generate regional metamorphism. The carnage I mentioned before is expressed as heat and pressure. So in class we will talk about the link between metamorphism and this type of plate to plate interaction. In Figure 7.10 we can see an example of plate tectonics at work. These mountains are on the island of South Georgia in the southern Atlantic. These mountains, now glacially carved, were produced by plate boundary interactions along the coast of South America, and do the continental movement, have not been located in the middle of the southern Atlantic.

Convergent boundaries are where plates come together. There are two types of plates, continental and oceanic, so that allows for three types of convergent boundaries. These are ocean to

Figure 7.10 **Southern Atlantic islands.**

ocean, continent to continent, and ocean to continent. Each one of these will have specific characteristics and geomorphological results. If we look at Figure 7.11, we will now focus on letter B. This is showing a simple diagram of a convergent boundary between ocean and continental crust. We call this

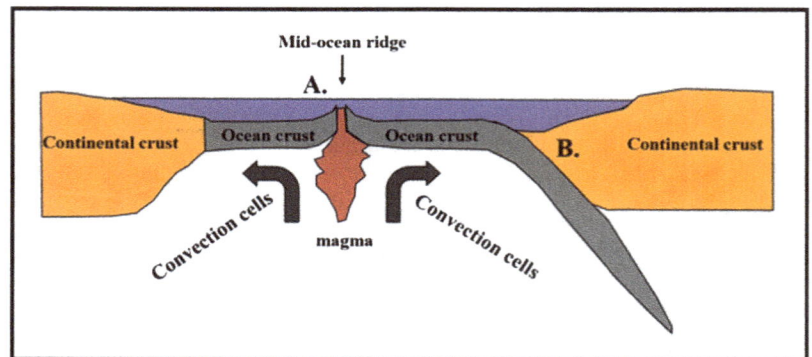

Figure 7.11 **Mid-ocean ridge formation.**

boundary configuration a subduction zone. Subduction just means that one of the plates is being pushed under the other one. Which one will be pushed under? The plate that has the highest density will be the plate that is pushed under and down into the Earth. Ocean curst is a mafic material whereas continental crust is a felsic material. You should remember from earlier chapters that mafic material is denser. This is why continental crust will ride above oceanic crust in subduction zones. The rule of density will density will help you remember which one will be subducted.

This works for ocean to continent, but what about convergent boundaries where there is not clear denser plate? The other two boundary types, ocean to ocean and continent to continent, are also easy to figure out. In the case of ocean to ocean, the plate that is denser will be subducted. How do we know which plate that is? We look for the plate that is older. The older the crustal material the colder the material will be. You remember from basic physical science that cold material will be denser than warmer material, so in the case of ocean to ocean, the older plate will be subducted. In continent to continent looking for the oldest plate does not really work. Both being intermediate to felsic in composition, neither plate will be dense enough to be subducted. The result will be both plates go up. This interaction can produce mountains of extreme height. You should be thinking of the Himalayas, but remember our humble Appalachians use to be just as high. They were both generated in the same way.

So now that we have worked out that ocean to ocean and ocean to continent will have a subduction zone, and continent to continent will not, we need to talk about what happens to the plate that is being subducted. Well, what do you think? You are taking solid rock and driving it deep down into the mantle where pressures and temperatures will be rising. Yes, you are correct in thinking that it will be metamorphosed, but it does not stop there. It will continue down until the material melts. When the material melts it will become more buoyant and will rise to the surface. So what do we call it when molten material rises to the surface? A volcano! So *ocean to ocean* and *continent to ocean* will produce volcanic activity whereas *continent to continent* generally does not. Let's take a look at Figure 7.12 showing these convergent boundaries.

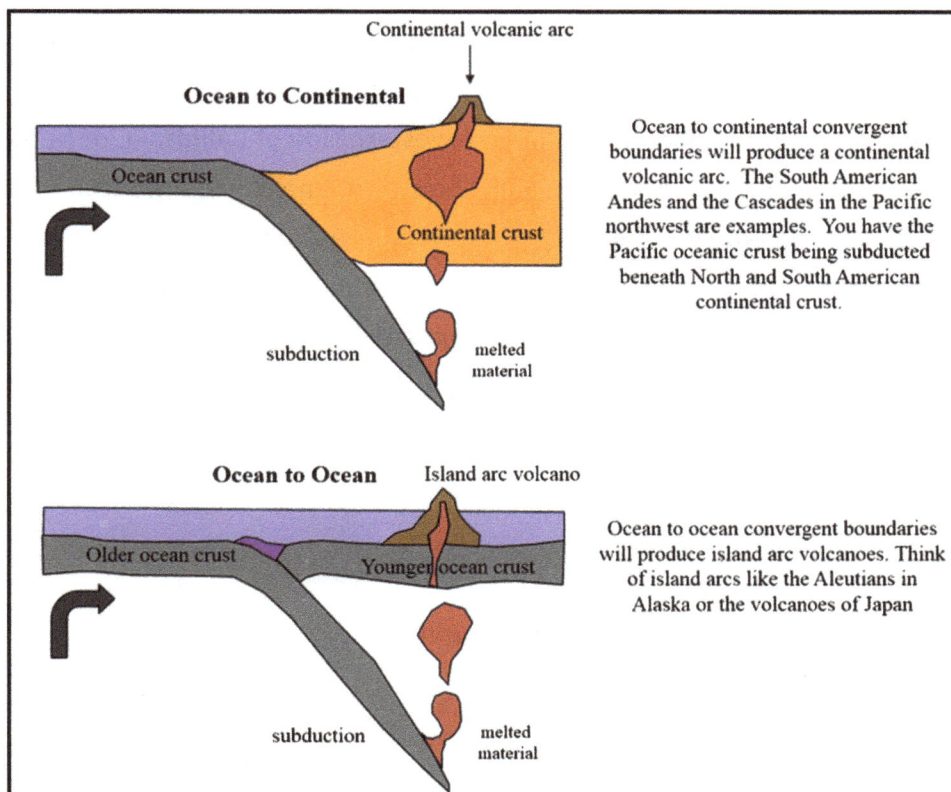

Figure 7.12 Continental and oceanic volcano development.

Additional assignment

Bonus Opportunity

Look up the term obduction. How does this relate to subduction? Draw a sketch of how this might happen. List several places on the globe where you might find obducted crust. What is the name we use to describe a section of ocean crust that has been obducted?

It is important to note the different types of volcanic activity associated with the different types of convergent boundaries. Make sure you understand the correlation between crustal type and the various volcanic arc types that can be created. Do not confuse this volcanic/ tectonic process with the one that produced volcanic islands like Hawaii. That is a different process that we will discuss later in this chapter.

Transform Boundaries

Transform boundaries are not as glamorous as the divergent or convergent boundary types. In the transform boundaries the plates are sliding past each other. We will refer to this as a strike-slip type of motion. This motion will generate heat due to friction, so there can be come types of dynamic metamorphism observed with this boundary type. The most common example is the transform boundary that outcrops along the western side of California, or better known at the San Andreas Fault system. In this particular boundary, the North American plate is sliding past the Pacific plate. This type of strike slip motion is needed to accommodate the combined motion of all the plates. In class we will look at some smaller transform boundaries and how they serve a function associated with divergent boundaries. There are no volcanic or mountain building events associated with this type of boundary. Earthquakes and dynamic metamorphism will be the primary characteristics.

Concept Review

1. What are the three types of plate boundaries?

2. What type of plate is denser? WHY?

3. How is density important to subduction?

4. What plate boundary type will produce island arc volcanoes?

5. Mount St. Helens would be an example of a volcano generated by what type of boundary?

6. Describe a rift valley?

7. Where can you find a currently active rift system?

8. What do the magnetic stripes on the sea floor represent?

9. What were Wegner's five pieces of evidence?

10. Why did no one believe him?

Hot Spots

Hot spots are covered in the plate tectonic chapter because they fall in a category in-between plate tectonics and volcanoes. Hot spots are exactly what the name states. These are sites around the globe where mantle plumes that come from deep within the Earth burn their way through both oceanic and continental crust. It is unknown what causes the eruption of this deep-seated mantle material, but we do know once they begin, they can be active for millions and millions of years.

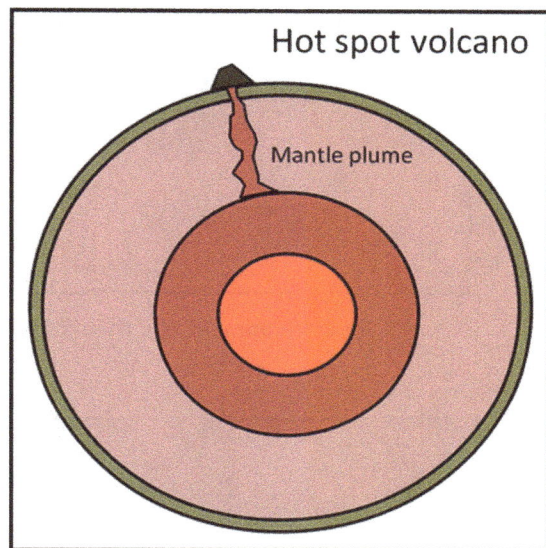

Figure 7.13 **Hot spot volcano.**

There are two examples that most people are familiar with, those are Yellow Stone NP and the state of Hawaii. There are many other island examples of hotspots and we will look at a few of them in lecture. The interesting feature about hot spot activity is that due to the plates being in constant motion, we tend to see long chains of mountains. Imagine holding a lighter below a sheet of paper. If you hold it in one spot, you will burn a hole. Slowly move the paper across the lighter, as a plate would move over the hot spot and you will burn many holes. This is why we see many volcanoes in a chain. It is important to note that there might be many volcanoes in the chain, but only the ones over the hotspot are active. Once the plate moves that volcanic center away from the plume, that volcano becomes extinct. Much like all the Hawaiian Islands except for the big island (Hawaii).

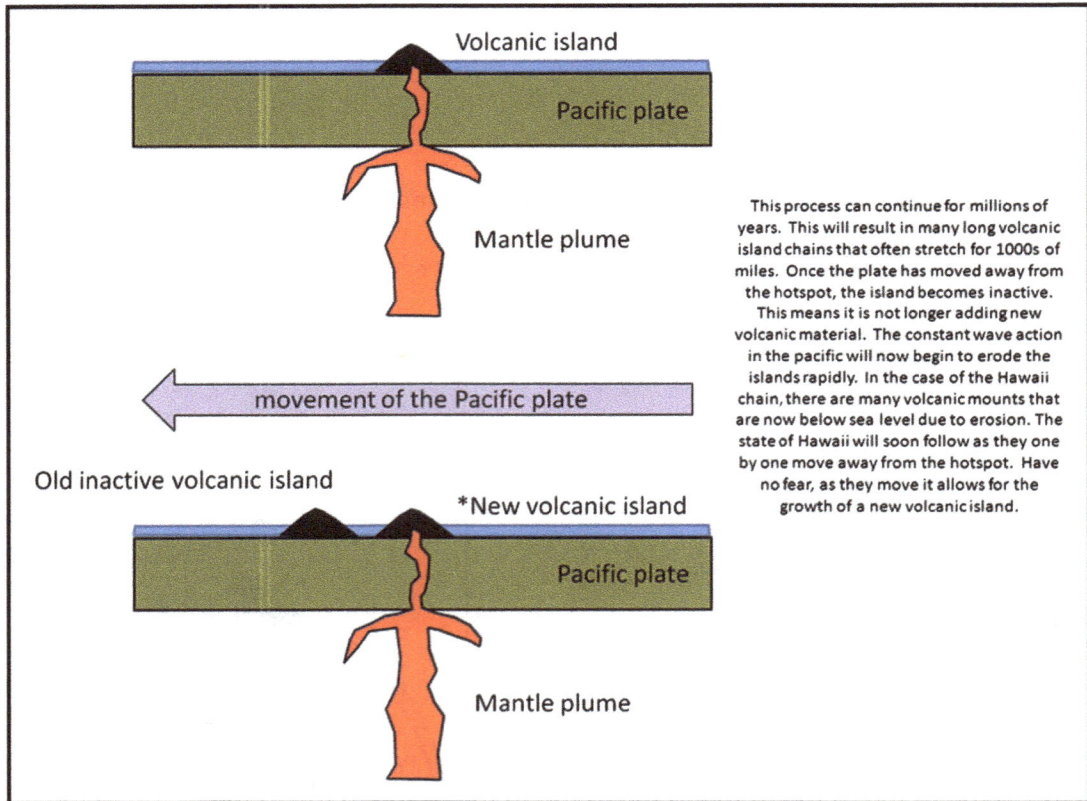

This process can continue for millions of years. This will result in many long volcanic island chains that often stretch for 1000s of miles. Once the plate has moved away from the hotspot, the island becomes inactive. This means it is not longer adding new volcanic material. The constant wave action in the pacific will now begin to erode the islands rapidly. In the case of the Hawaii chain, there are many volcanic mounts that are now below sea level due to erosion. The state of Hawaii will soon follow as they one by one move away from the hotspot. Have no fear, as they move it allows for the growth of a new volcanic island.

Figure 7.14 **Development of a chain of volcanic islands.**

Figure 7.15 **Steep shores of big island of Hawaii.**

As you can see in Figure 7.15, the shores of the big island of Hawaii are rather steep in places. This is due to the constant wave action acting upon the soft volcanic rocks. Over time the rain and other erosional processes will destroy the exposed part of this island. Currently the volcanic activity is occurring on the east edge of the big island. Speaking in geologic terms, the big island will soon move off of the hot spot.

Concept Check

How do hot spot volcanoes form? Now compare and contrast this with oceanic arc and continental arc volcanism.

CHAPTER 8

STRUCTURAL GEOLOGY:
CRUSTAL DEFORMATION AND MOUNTAIN-BUILDING

Structural Geology

Structural geology is the study of rock orientations and deformations in the attempt to understand the past forces that have generated current orientations and deformations. This information is used to unravel how mountains are formed, how continents break apart, and can explain why the surface is the way we see it today. Structural geology has many uses in the academic world as well as the commercial one. Many structural geologists are responsible for following a gold vein or petroleum deposit. Having a basic understanding of structural orientation and deformation will help many students see the world around them in entirely different ways.

Deformation

Deformation is a change in shape and/or volume of a rock.

In order to talk about structure, we need to talk about deformation. The above definition of deformation refers to the shape and volume of rock material. We will look at both of these types of deformation individually. But before we get into that, I want to talk briefly about the mechanism behind this deformation. In order to build mountains, create valleys, and deform rocks, there must be a powerful energy source. This source is plate tectonics. When we talk about crustal deformation, keep in mind the different boundary types we learned about in plate tectonics, and remember that building mountains is a messy process. It will deform large amounts of crust.

When speaking of deformation, we talk about changes in shape and/or volume. When we have strictly just a shape change, we call that distortion. **Distortion** is a change in shape, but volume remains constant.

All images in this chapter are courtesy of John Hawkins

We can see that distortion only changes the shape, but what if we do change the volume? In case of a volume change, we call it dilation. **Dilation** is a change in volume.

So now, when we talk about deformation, we can identify it in terms of distortion and dilation. It is importation to keep in mind that during deformation, it is common for both distortion and dilation to occur on the same material at the same time. These forms of deformation can also be referred to as strain. This now brings us to two terms that are often confused. These are **Stress** and **Strain**. Put simply, stress is the force, the event, the energy, the action, that causes the strain. The strain is the result of this stress, the observed effects, or the damaged result. In geology, we often do not get to observe the actual stress on the material; we are only able to observe the effects of stress as strain.

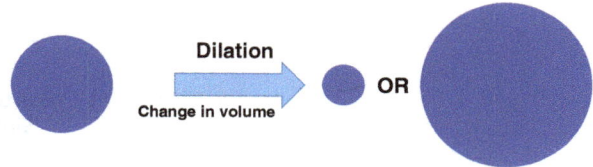

Stress

We have four main types of stress. As we go over each type, think about plate tectonic boundary types. You should be able to relate the first three types of stress back to a specific plate tectonic boundary type.

1. **Tensional:** Pulling apart; thins and extends the material

2. **Compressional:** Pushing together; thickens and shortens the material.

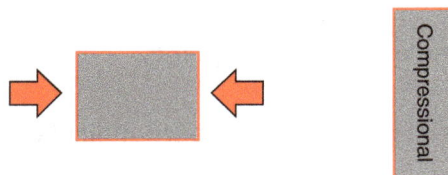

3. **Shear:** Rotational stress, clockwise or counter-clockwise.

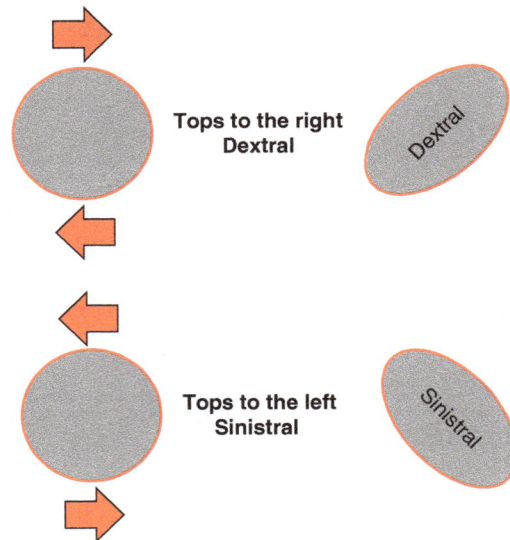

Tops to the right
Dextral

Dextral

Tops to the left
Sinistral

Sinistral

4. **Hydrostatic:** Stress comes from all sides, confining pressure.

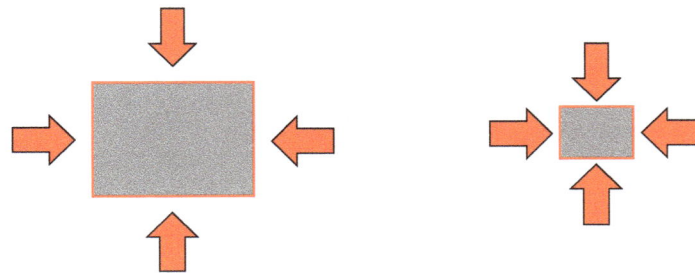

These are the four types of stresses that will produce strain in our rocks. One common by-product of stress is the production of foliation (remember the pressure section of metamorphic rocks), and the tilting of sedimentary rock strata. This tilting can affect any rock type, but in order to discuss our next topic, let us keep things simple. So, let us imagine that stress has affected some old sedimentary rocks. These rock have become tilted over time. In order for a geologist to record and describe this tilting, we need a system to discuss the orientation. We call this system "Strike and Dip."

Strike and Dip

Strike: Bearing of a horizontal line on an inclined plane.

Dip: Maximum inclination of the plane, measured perpendicular to strike.

Now, I know that you just read those two definitions and probably have no idea what they are talking about. That is okay. The idea of strike and dip can be challenging. It usually becomes a much easier concept when outlined visually. We will have several examples in class, and even if this concept is unfamiliar to you, do not worry. Many of our geology majors do not fully understand strike and dip until later in their coursework. So, we are going to keep it simple and work on a functional understanding of strike and dip.

Let us first talk about strike. Look at Figure 8.1, you can clearly see an outcrop sticking up out of the sand and it is mostly covered in seaweed. These are tilted sedimentary rocks, and there-fore, we need to describe their orientation. Time to break down the definition of strike. First, bearing. What is a bearing? It is just the direction. Now, we cannot use terms such as left, right, or straight ahead because those are arbitrary. So, we will use our compass directions (North, South, East, and West) as the basis for our bearing. So, bearing is just going to be a compass direction. Second, horizontal line. A horizontal line is really easy to visualize. Now, imagine you jump up on that layer of tilted strata. You want to walk along those rocks and continue down along the beach. In order to do so, you must walk in a horizontal line on that inclined plane to continue down the beach. So, the strike will be the bearing of the direction you are walking in. The red arrow generally highlights the direction you would be traversing. If we measure the compass direction of that red arrow, we will have our strike. Third, inclined plane. The last part of this definition tells us that the rocks must be inclined. If the rocks are inclined. If the rocks are horizontal, then strike does not work. We will talk in class about what to do in the case of horizon-tal strata.

Dip is a much easier concept to grasp than strike. Earlier, you read the over-worded and complicated definition for dip. The easiest way

Figure 8.1 Image of tilted rock units along the coast of Scotland. The red arrow is showing the general direction of strike for these rocks.

to think about dip is this: If I am sitting on those rocks, and I spill my drink, which direction will the fluid flow? The answer is simple—It will flow downhill. The idea of downhill is a perfect way to think about dip since dip is the downhill direction. So, to find dip, we have to find the steepest downhill part of the inclined rocks. Gravity will really help out with this one. When in doubt, just spill your drink and the dip will be revealed. We would then note the compass direction and the magnitude of the incline to accurately describe the orientation. The interesting relationship between strike and dip is that they are always perpendicular to each other. If you can find dip, strike will be 90 degrees out from it. In class, we will talk about the strike and dip symbol. This is the symbol that you will see on all geologic maps and it helps geologists to understand the 3-dimensional orientation of the rock layers from a 2-dimensional paper map.

Concept Check

1. Relate each of the first three types of stress fields back to a plate tectonic boundary.

2. Explain the difference(s) between stress and strain.

3. What are the two types of deformation?

4. In paragraph form, please explain the concept of strike and dip, and why it is necessary for geologists.

CHAPTER 9

VOLCANOES
CREATIVE AND DESTRUCTIVE

Volcanoes. . .

Volcanoes have captured our fascination with their unmatched power to create and destroy for as long as we have known them to exist. Unless one has the unfortunate opportunity to be caught within an eruption, it is hard to fully explain this horrifying power. Many of us have very little experience with volcanoes, and what experience we do have has come from television shows or documentaries. The most often visited volcanic regions in the US are Mount St. Helens and Yellowstone National Park. The last major eruption in the lower 48 states occurred at Mount St. Helens on May 18, 1980, and no humans observed the last eruption of Yellowstone. So in our collective memories about volcanoes, the details are becoming muddled with each passing year. In order to learn more about this topic, I had the opportunity to attend an international volcanology field school in Kamchatka, Russia. This opportunity taught me that the topics we are about to cover in this course are broad brushed concepts at best. I learned that each volcano can be diverse and varied, thus making it difficult to place some into the typical 3 categories that are often used. In this course we will talk about these three categories and the typical characteristics of each. I do want you to remember

Figure 9.1 **Relaxing on a volcano.**

that no volcano is typical and its behavior can deviate from "typical characteristics" at any given moment.

Now before we break volcanoes down into their 3 assorted types, we are going to take a look at them with a more simplistic division. This division will be based on if the volcano is "quiet" or "explosive". When I use the term quiet, you should be thinking about a situation like the one found on Hawaii and in the Hawaii Volcanoes National Park. When I use the term explosive, you should think of Mount St. Helens or of Mount Vesuvius. There is one characteristic that helps to determine if a volcano will be quiet or explosive. That characteristics is the amount of silica found in the molten material that the volcano is erupting. If the material has a high silica content (more felsic in nature), this makes the material very viscus and allows it to trap a large quantity of dissolved gasses. It is these dissolved gas that cause pressure to build resulting in a volcanic eruption. On the other hand, if the material has a low viscosity (more mafic in nature), the dissolved gasses are allowed to escape through the volcanic vent and therefore pressure does not build. No pressure equals no explosion. So the bottom line in determining the broad brush safety factor of a volcano is to determine the viscosity/composition of the material being erupted.

Figure 9.2 In this image you can see the plume rising from the crater in Hawaii Volcanoes National Park. All of those volatiles (dissolved gasses) within the magma are easily escaping the vent. This is because the material in the Hawaiian volcano is mafic. Pressure is not allowed to build.

We have been mentioning erupted material. So what vocabulary terms will we use to discuss this material? You are familiar with the terms lava (molten rock erupted onto the Earth's surface) and magma (molten material still within the Earth's surface). Both lava and magma are present in the plumbing of all volcanic systems. Magma becomes lava and if lava flows back underground then it can become magma again. So these terms are generally used to describe the location of the molten material. When we hear the term lava, we generally think of black fields of harden material. That is not an incorrect assumption or mental image if you are thinking about what it would look like after the lava has cooled. However, lava will come in two main types. We call one **Aa**, and the other **Pahoehoe**.

Figure 9.3 a-b Aa and Pahoehoe lava.

The Aa variety of solidified lava is generally crumbly with sharp jagged edges. If you were to walk over this material barefoot you would probably make a sound very similar to AHHH-AHHHH. Pahoehoe lava will stand in stark contrast to Aa. You can see in Figure 9.3 that the pahoehoe is very smooth and tends to fold like fine fabric. You can see in Figure 9.3b that a student is examining the "ropy" folds and curves of the pahoehoe. Both of these flows were generated by the same volcanic vent. You can clearly see that the pahoehoe is on top of the aa, indicating that it is the most recent material that has been erupted.

Both of these lava types will be between 1000-1200 degrees Celsius and can move at variable rates. The aa flow will move like a tank tread. It will be crumbly on the top and bottom with molten material in the center. The rate of movement, as with any flow, will depend on the slope gradient and the pressure of the flow behind it. Both aa and pahoehoe can move rapidly or slowly. However, generally speaking, aa will crumble forward more rapidly than the slow silky flow of the pahoehoe.

Figure 9.4 **In the above image, you can see many different flow pulses that have occurred over a 20 year period. This is a volcano in Russia, and it is very different from the volcano found on Hawaii. You can clearly see the rubbly/blocky nature to the flows on the flank of this volcano. Aa and pahoehoe can be found on volcanoes of different types, as long as the magma/lava composition is more mafic. We will talk about the composition of magma/lava and its overall effects in class.**

Types of Volcanoes

There are three main types of volcanoes that we will cover in this chapter. Volcanoes are differentiated based on their overall geomorphology, lava/magma composition, and ejected material. We will talk about each of the three listed below and see how they compare based on these three criteria.

- **SHIELD**
- **COMPOSITE**
- **CINDER CONE**

Shield Volcanoes

Shield volcanoes are among the largest mountains in the world. The best examples of shield volcanoes are the two that are present on the big island of Hawaii, Mauna Kea and Mauna Loa. Mauna Kea stands 13,802 feet above sea level. When you factor in the depth to the bottom of the ocean, this makes the big island of Hawaii the tallest mountain in the world. Not the highest, but it is the tallest. We will talk about the formation of this volcano in more detail in the plate tectonics chapter, but it is feed by magma that is rising deep from the mantle. The composition is *primarily* **mafic** and these volcanoes are geomorphologically characterized by gentle sloping sides. The slopes will range from *~2degrees at the base to ~10 degrees* at the summit. In Figure 9.5, you can see the gentle slopping sides coming down from the central cone on Mauna Loa. The material that is ejected from the central cone is almost primarily flowing lava of both aa and pahoehoe varieties. These flows can and have stretched for miles down the side of the mountain. This does not mean that this volcanic system cannot erupt other materials that we will discuss in this chapter, recall that we are speaking in generalizations. However, due to

Figure 9.5 View of Mauna Loa in the distance taken from the top on Mauna Kea on the big island of Hawaii.

the ever-changing position of the Hawaiian Islands, Mauna Loa is slowly moving away from the deep seated magma plume that feeds this volcanic system. If you visit the big island, you will notice that most volcanic activity has now moved more to the west of this past volcanic center. The peak of Mauna Kea has been deemed safe enough to construct large telescopes as seen in Figure 9.6. Once the magma center has moved from the hot spot beneath it, the volcano will no longer be supplied with eruptible material.

Shield volcano key points-

- Gently sloping flanks (~2 to ~10 degrees)
- Largest type of volcano with respect to area

Figure 9.6 **View of the Keck telescopes on top on Mauna Kea with some small cinder cones to left center.**

- Primarily mafic in composition

- Aa and Pahoehoe lavas are the primary product

- Mauna Kea and Mauna Lou are examples

Composite Volcanoes

Composite volcanoes are generally the idea that most people will conjure up when they hear the word volcano. Most people do have an understanding of this type either due to their popularity in the movies or to stories of past devastation that composite type volcanoes can cause. Composite volcanoes can also be referred to as "Strato" or "Layer-cake volcanoes". One of the best examples of this type of volcano is Mount St. Helens. In Figure 9.7, you can see a similar volcano to that of Mount St. Helens, Mount Rainer. This image was taken from an airplane on approach to Seattle. It does lack perspective, but these volcanoes can reach heights of 8-10 thousand feet. This is height above the Earth's surface and should not be assumed to be their height above sea level. Looking at this type geomorphologically, they fit the standard idea of a volcano. They have steeper sides than a shield cone, reaching up to ~30 degrees at the summit and level off to near ~15 degrees on the flanks. In class we will talk about how composite volcanoes will differ from shield volcanoes in composition. If you recall, shield volcanoes are primarily mafic, and by contrast we teach that composite volcanoes are more intermediate in composition. Is this an incorrect statement? No. Composite volcanoes are, for the most part, intermediate in composition. It is the additional silica in the magma chemistry that allows for this different shape and a different eruptive characteristic. The gas/volatiles that we discussed that were freely leaving the vent of our shield volcano are now becoming trapped and will escape in a series of more

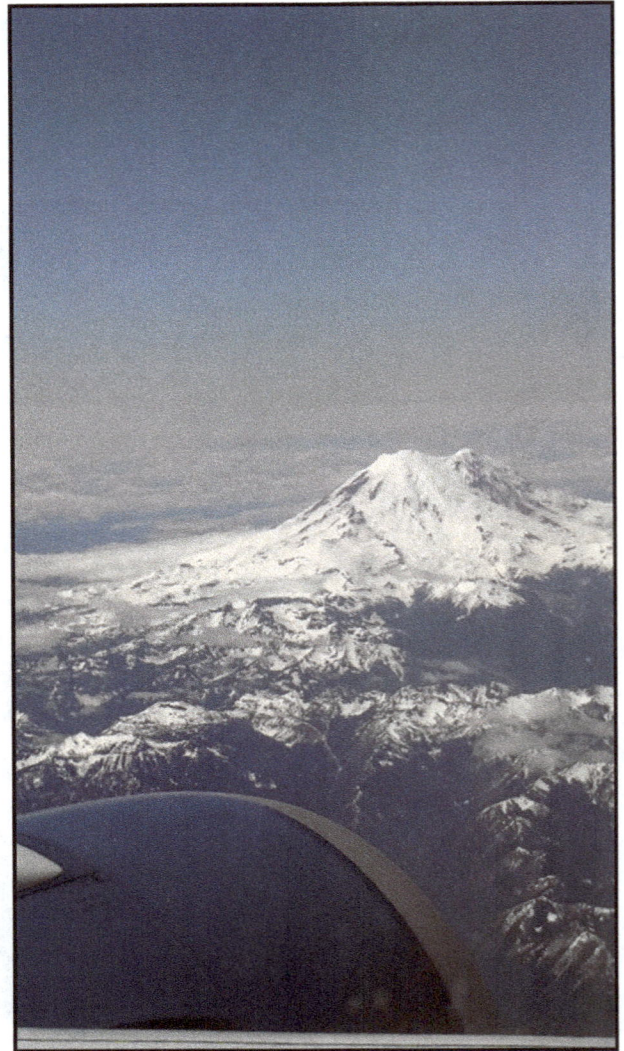

Figure 9.7 Mount Ranier.

explosive periods where we get ash as the main eruptive material rather than lava. These volcanoes will still produce lava, but now it is combined with layers of ash. It is these alternating layers (ash,lava,ash,lava,ash,lava) that give this volcano its name. Composite, Strato, and Layer-cake are all referring to the layered make up of the cone.

Getting back to the discussion on magma chemistry, let's take a look at the volcano in Figure 9.8. You can clearly see the slope angles match our geomorphologic description. I will also tell you that this volcano can and does produce large amounts of ash and lava flows. I had to deal with the ash from this volcano getting into my food and sleeping bag. So trust me when I say, "It produces lots of ash." It is a composite volcano; however, it is *mafic* in composition. So when we say that composite volcanoes are intermediate, that is mostly the truth. I just need you to understand that in geology there are always exceptions, and one should never jump to make assumptions.

The added silica in these volcanic systems will make them much more explosive. Imagine shaking up a can of coke and then popping the top, the dissolved gassed escape rapidly and turn the liquid material to foam. This happens in volcanoes

Figure 9.8 A composite volcano, however it is mafic in composition.

Figure 9.9 a-b This is the Russian field house that we were housed in during our volcanic expedition. There are no roads to this location, only way in is by helicopter or snow mobile in the winter. Building this station was no easy task. The structure is insulated by volcanic material called pumice. We had to sweep out the ash from the house each morning.

as well. When the dissolved gasses increase in concentration, it will result in a buildup of pressure. When this pressure is released, large quantities of hot ash and volatiles will be expelled as a fiery could. We call this cloud a *Nuees Ardentes* or a *pyroclastic flow*. These clouds are devastating to anyone or anything caught in their path. The cloud will have temperatures of 800°C and above, so it will glow red in the dark. It will hover across the ground and incinerate anything that is passes over. This hover effect allows the clouds to travel quickly; they have been clocked at speeds near 100 mph. We will look at some examples in class and talk about how dangerous these pyroclastic flows really are, and we will look at some examples of cites that have been over taken by this process. I do not have any images that I have taken of pyroclastic flows, but I have worked with older ones. In Figure 9.10, you can see an individual walking on a slope right below a layer of larger sized material. You can clearly see the finer grained ash below his feet. This finer grained material

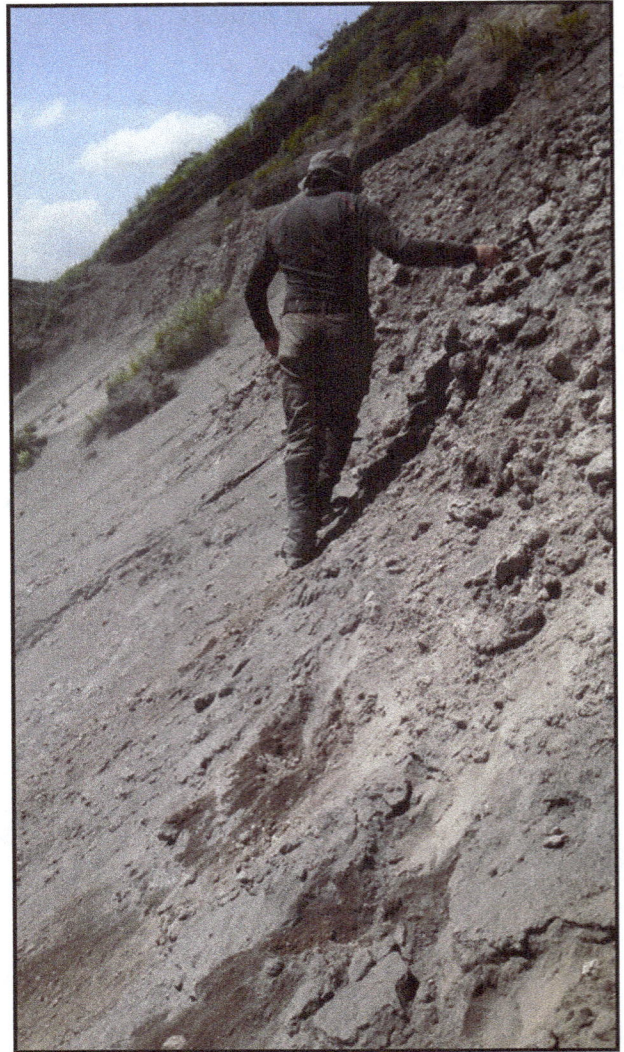

Figure 9.10 Walking on a slope with finer and larger grained materials.

is older and is the same type of material found covering the entire volcanic area. However, this portion of fine grained ash was covered by a pyroclastic flow. The coarser grained material above was what was carried within the pyroclastic cloud. This would have been violently and rapidly deposited.

In addition to ash, we can see above that ejected material can be rather large. We call these ejected materials volcanic bombs. These can range in size from the size of a quarter to upwards of a small SUV. This material will rain down from the volcanic cone and are often one of the most dangerous aspects to an eruption. It is impossible to predict where these small pieces will land, and they can travel several miles from the central cone.

Another example can be seen in the drop stone/volcanic bomb (Figure 9.11). This is an image from the coast of Scotland from during a time when it was very volcanically active. Even though that was millions of years ago, we can still see the evidence along the beaches. In this image you can see a small (baseball sized) volcanic bomb.

Figure 9.11   Drop stone/volcanic bomb.

The term drop stone was used by the geologists from St. Andrews in Scotland, so we will use those two terms interchangeably. You can clearly see in the photo how the heavier drop stone landed on top of several layers of finer grained ash. It deformed the layers upon impact. This is excellent evidence to suggest this piece was airborne and landed with significant force.

Volcanic Ejecta

In image A we can see a large bomb that had been ejected from the central vent. This more than likely tumbled down the volcanic flank, but it would still be rather destructive. In image B we see what is called a "bread crusted" bomb. It takes its name from the outer edge with its crusty cooling texture. Many liken it to crusty bread, hence the name. In this image you can see the tip of a walking stick for scale. This bomb was found over 2 miles from the volcanic vent, so this would have been a projectile during some previous eruption. In Image C we can see in inside of a piece of ejected material. You will notice the vesicular texture that we have discussed in previous chapters. This is a sample of pumice. Ejected volcanic material can have a vesicular texture or you can have them that are more massive. In picture D you can see a comparison of a volatile rich pumice material next to a sample that had completely out gassed before it had been erupted. If material is ejected rapidly, then you will generally see a vesicular texture, however if the molten material is allowed to experience lower pressures before eruption, then it can allow its dissolved gasses to escape before it hardens during the eruption. This out gassing will produce material that is denser than the floating pumice that we have discussed before.

Volcanic Inheritance

There is one other idea that I would like to discuss during this chapter. This idea has to do with volcanic inheritance. The use of the term inheritance here basically means the same thing as it would when speaking about your heritage. In this case it is referencing to a volcanic centers tendency to form in areas of past volcanic activity. For example, let's say we have a composite volcano that has been erupting for 1000s of years. On occasion, volcanoes can blow themselves out of existence. Completely destroying the cone and leaving a large caldera/crater where the volcano once was. With the idea of inheritance, this means that it is likely that a new volcano will grow close to or in the same location as the one that just exploded. Figure 9.12, we can see a crater in the back of the image. This was a volcano that exploded long ago. In the foreground you can see the black flows coming from the recent volcano named Karymsky, and it appears that these lavas are hitting the side wall of a crater. That is because the current volcano Karymsky is sitting inside the caldera/crater of a volcano that exploded before Karymsky started to form.

Figure 9.12 Crater in background from old volcano that exploded long ago. Foreground shows black flows from recent volcano Karymsky.

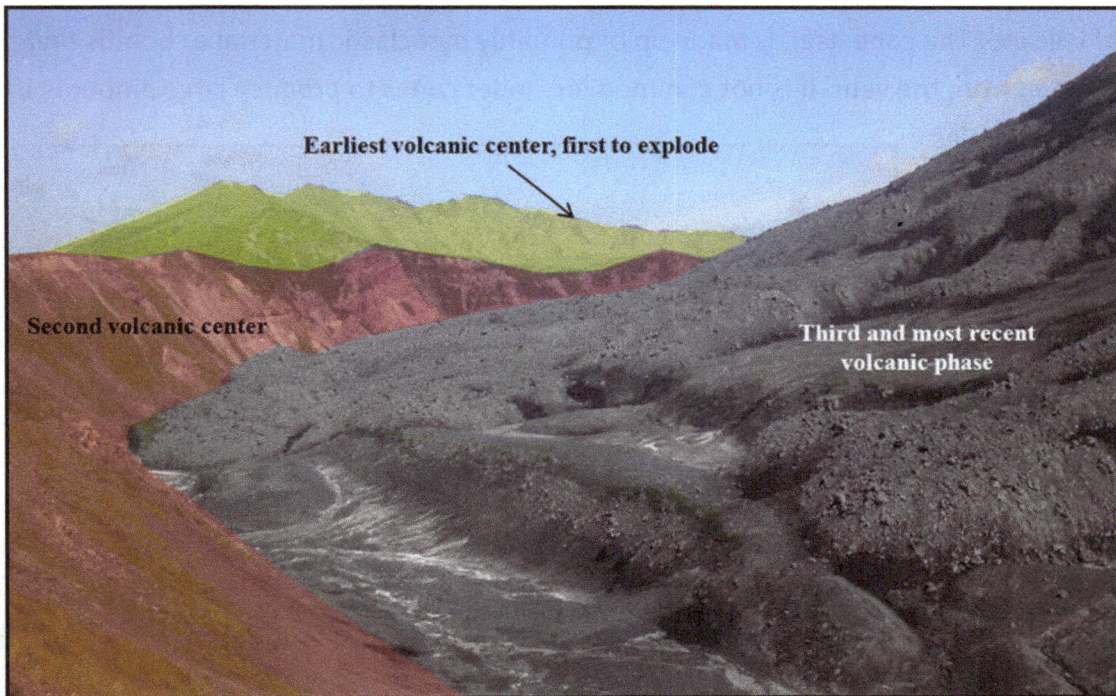

Figure 9.13 Colored screens distinguish volcanos.

So in this image we have 2 former volcanoes that have blown themselves out of existence, but due to volcanic inheritance, we have one currently in the same location. One day it is likely that Karymsky will explode, and it would be safe to assume a new volcano would form in its place.

Composite Volcano key points-

- Typically ~8,000 to ~10,000 feet high

- Classic volcano shape, (15° along flanks ~30° near the vent)

- Mostly intermediate compositions

- lava flows mixed with pyroclastic eruptions

- example, Mt. St. Helens and Mount Vesuvius

Cinder Cones

The third volcanic type that we will cover in this chapter will be **Cinder Cones**. These are generally smaller than the previous two types by at least an order of magnitude. They are geomorphologically characterized by steep sides ~30 to ~45 degrees and seldom reach heights much above 500 feet from the surface. So you can see why these are considered to be the smallest type of volcano. The cone itself is made up of primarily pyroclastic material or bombs that have been ejected from the vent. It is not common for cinder cones to produce large amounts of lava. There are examples where lava flows have erupted from the base of the cinder cones, but seldom from the vent. This is what separates this type from that of a composite type volcano and these are generally found adjacent to either composite or shield volcanoes. In Figure 9.14, you can

Figure 9.14 Four cinder cones on top of Mauna Kea in Hawaii.

Figure 9.15 a-c The photograph to the left was taken from the top of Capulin Volcano National Monument in southern Colorado. From the top you can see the ancient lava flows that once erupted from the base of the cinder cones. Once cinder cones stop erupting, it does not take nature long before it will reclaim the space.

make out 4 small cinder comes that are sitting on top of Mauna Kea in Hawaii. This will hopefully help you understand the difference in scale when it comes to the three types of volcanoes.

Cinder Cone key points-

- Smallest type of volcano

- Steep, straight-sided slopes (30°- 40°)

- Exclusively pyroclastic material

Hydrothermal Events – Geysir/Geyser

In most volcanic areas you will also find hydrothermal vents or geysers. These are very common in the Yellowstone area and draw many tourist a year. These occur due to the ground water coming into contact with rocks that have been heated due to magma in the subsurface.

The ground water flows along until it comes into contact with the hot material. The water will become heated till it begins to boil, and the boiling action will force the liquid to the surface. The resulting fluid expulsion is what we call a geyser. The most common geyser in the US is "Old Faithful". The reason that this geyser is on a reliable schedule is that the rate at which the ground water will refill the void left after the previous expulsion is fairly consistent. During years of drought or after earthquakes, this rate of water flow can be slowed or disrupted, and that would change the schedule of eruption. The word geyser comes from the Icelandic word geysir. In the image below you can see the type locality for geysers in Iceland.

In addition to hot fluids there are also dissolved minerals that erupt from geysers. In many cases various types of limestone, called travertine, can be formed. In addition to limestone, many geyser fluids can be silica rich. If you look near the outflow of the geyser you will notice silica gels forming in the shapes of small needles or as small terraces. In Figure 9.16 we can see an example of silica gel accumulating in terraces from an active geyser in Russia. All volcano types can have active hydrothermal vents. In some cases, there is not enough water in the water table to produce eruptive geysers, and this will result in steam vents. Figure 9.16d you can see one of our students standing in a steam vent found on the big island of Hawaii. Volcanoes are constantly adding gasses to our atmosphere, with a majority of this vapor being water vapor. It is theorized that in Earth's early history, volcanoes were a major contributor of water vapor to the surface. In addition to water vapor, there will be sulfur rich gasses notably sulfur dioxide, and carbon dioxide. When visiting volcanic areas it is always important to stay upwind from outgassing vents. Concentrations of carbon dioxide and sulfur dioxide can reach poisonous levels easily.

Being a geologist, it is our job to travel to these wonderful places and study this natural beauty. Often we are sent into places that have yet to be properly mapped or explored. So even in 2015, you can still be an explorer and travel to remote places that most humans will never see. When you become a geologist, the planet becomes your cubicle. Below are just some images as an example of some of the wonders that geology can create. If you have any questions about becoming a geologist and making the choice to have an adventurous life, feel free to ask.

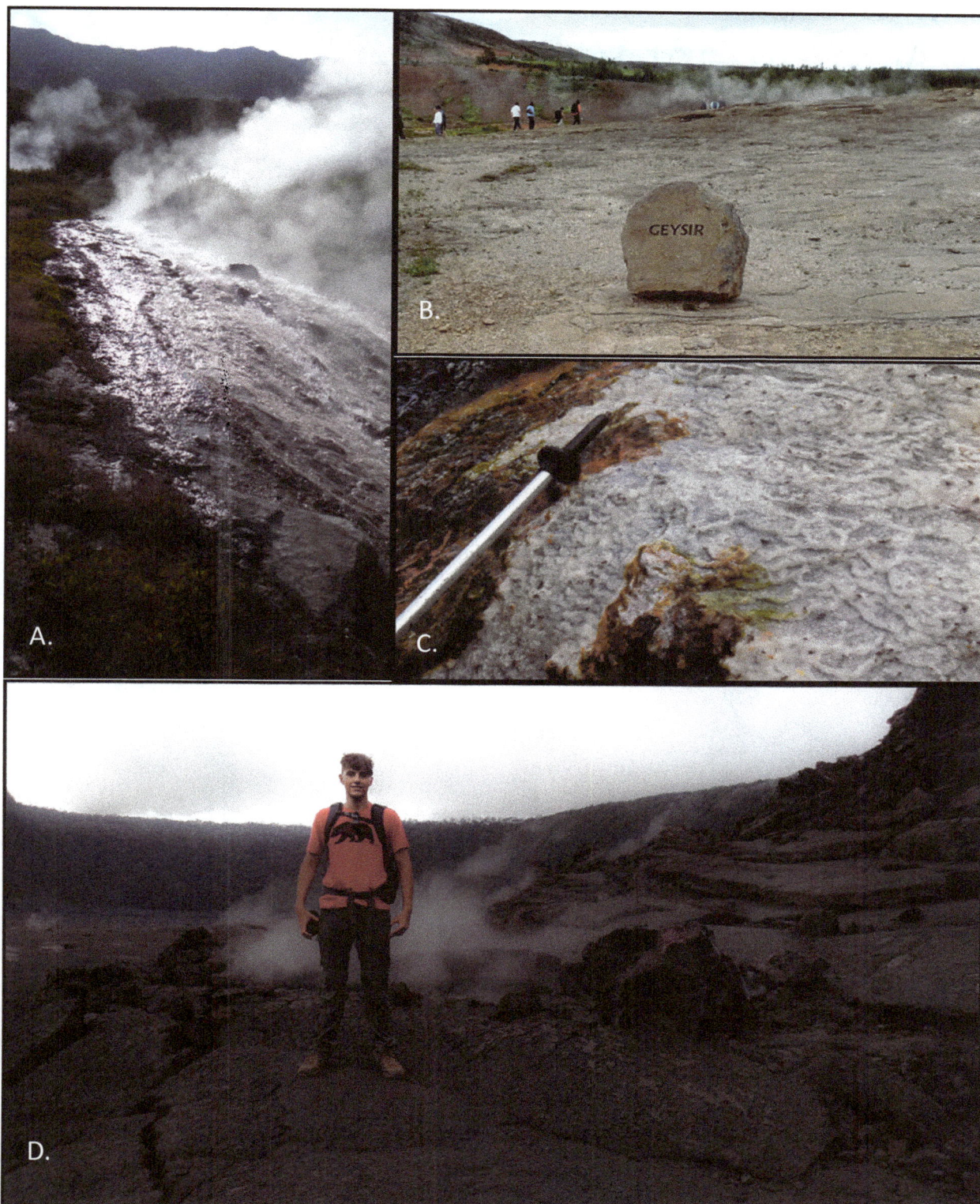

Figure 9.16 a-d In photograph A, we have a geyser in a volcanic filed in Russia erupting vapor and silica rich fluid. The silica gels seen in photograph C, were found in the waste waters of this particular geyser. In photograph B, we see the sign for the type location of the word geyser in its Icelandic spelling. In photograph D, we see one of our undergraduate students standing in the vapor eruption from a steam geyser in a recently cooled caldera in Hawaii.

Monument Valley

Grand Canyon

Russian Volcanic Field

Arctic Norway

Scottish Highlands

Antarctica

CHAPTER 10

GLACIERS:
PROCESSES AND LANDFORMS

What are Glaciers?

This topic is rapidly becoming more important to the mainstream public. It is likely that in your adult life, there will be many tough decisions that will have to be made in response to what might be happening to glaciers worldwide. In order to fully understand what is now inevitable, we need to make sure that we have a basic understanding of glaciers and the associated glacial processes. In this chapter, we will talk about these processes and the various types of landforms associated with glacial movement and meltwaters.

Since most of us are from the southern U.S.A., it is not easy to gain firsthand experience with glaciers. Either you go on that Alaskan cruise with your grandparents or rely on what we see on TV or the Internet. Luckily, however, glaciers are similar to something that we are familiar with—ICE. Now, in spite of what many will say in the upcoming years, ice does not have a political agenda. It does not care about your opinions, or anything else,

for that matter. Ice just melts. Believe it or not, ice has always melted. It becomes a concern when the rate of melting is faster than the natural processes can replace it.

All images in this chapter are courtesy of John Hawkins

This is probably what most of us would consider as the natural habit of ice. When we talk about glaciers and glacial ice, we are talking about a substance that is similar, yet different from the stuff you will find in your local gas station.

So, let us start with a simple observation—Glaciers are defined by the presence of year-round ice. Ice/snow that falls during one winter that does not melt and that will become covered by the coming year's ice/snow is the start of the glacial process. This process is slow. We cannot have one cold winter and expect to restore the glacial mass that we have lost in the past 20–50 years. This process takes multiple years of year-round ice. It is this compounding ice that allows for pressures needed to change the snow into glacial ice.

Figure 10.1 **In this image, you can see the alternating bands of yearly snowfall. Now, imagine as these accumulate slowly, year after year.**

In Figure 10.1, you can see multiple layers of lighter and darker bands. These bands represent past years of snowfall. So, you can see how this accumulation will lead to deeper pressures below. Remember—Ice is a mineral! In this process of glacial ice formation, we are changing the mineral into a different crystal structure by adding pressure. This is, in effect, a kind of metamorphism. So, glacial ice is metamorphosed snow that fell over many years in the past. So, hopefully, now you will understand why the process of formation of glaciers is a slow one.

This process leads to the two broad categories of glaciers: Alpine and Continental. Alpine glaciers are the ones that most of us are familiar with. These are the ones we will see in higher latitudes or higher elevations that are within mountain ranges. You can typically think of Alaska when you think of a location that has lots of alpine glaciers. There are several states in the Western Lower 48 that have alpine glaciers. In Figure 10.2, we see an alpine glacier located in northern California. This glacier, the Ritter Glacier, is located ~12,000 feet above sea level. It is the cooler temperatures at this altitude that allows the glacier to survive. Other glacial examples exist along the Pacific Northwest. In Figure 10.3, we can see multiple alpine glaciers clinging to the side of Mt. Adams in the Cascade Mountain range. We will take a closer look at the alpine glacial landforms a bit later in this chapter.

Figure 10.2 This is the Ritter Glacier found in the Ritter Range of Northern California. Image taken in June 2008 at an Elevation of 12,000 feet.

Figure 10.3   Mt. Adams as seen when flying into Seattle. Image taken in May 2016.

In addition to the glaciers we see today, during the last glacial maximum of 18,000 years ago, many locations that you would not expect to have glaciers did. Hawaii, for example. Keep in mind that the summit of Mauna Kea (Figure 10.4) is over 14,000 feet above sea level. You might associate Hawaii with beaches and warm temperatures, but altitude can make a lot of difference.

Continental glaciers are harder to visit than their alpine counterparts. During recent times, continental glaciers are only to be found in Greenland and Antarctica. Continental glaciers use to cover one-third of the globe roughly 18,000 years ago. During this time, much of our landmasses in higher latitudes were scoured down to the bedrock and many of the landforms we see today are a result of these glacial advances. Keep in mind that continental glaciers can range from hundreds of feet to several miles in thickness. It is the large masses that can generate large icebergs the size of city blocks (Figure 10.5).

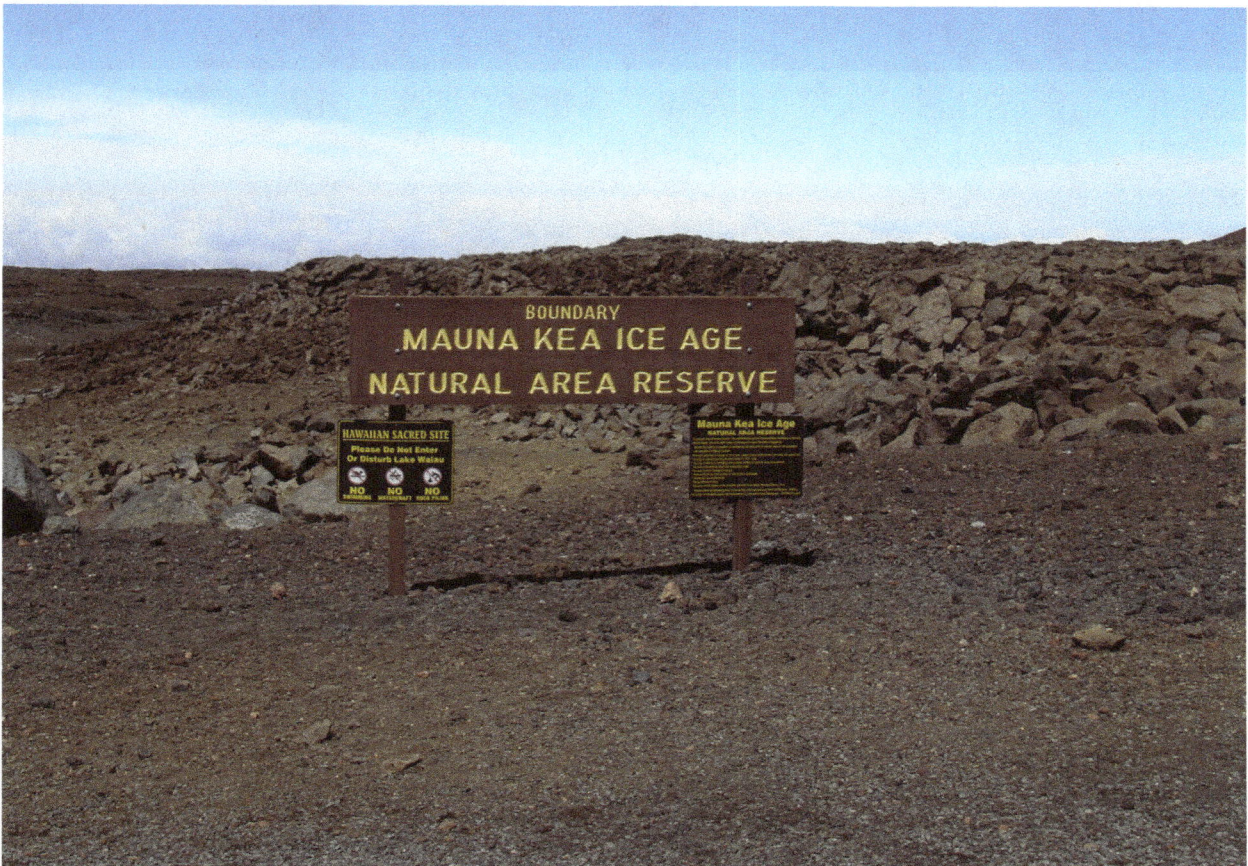

Figure 10.4 **Summit of Mauna Kea on the Island of Hawaii. Image was taken in May of 2016**

Figure 10.5 A. Image of the leading edge of a continental glacier. You can see the 15-passenger zodiac in the forefront for scale. Image taken along the coast of Antarctica. B. Image of large icebergs that broke off from continental glacier ice sheets. Some of these icebergs are the size of city blocks.

Glacial Anatomy

In this section, we are going to keep glacial anatomy simple. We are going to focus on three parts of the glacier: Zone of Accumulation, Zone of Ablation, and the Equilibrium line. These

Zone of Accumulation

Zone of Ablation

Equilibrium line

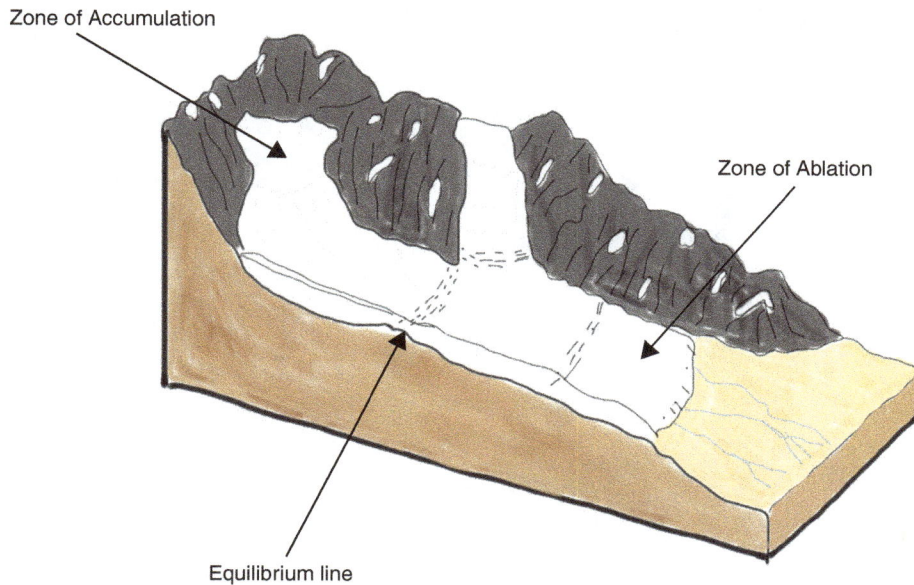

three sections are much easier to view on alpine glaciers. The following diagram will illustrate these three sections and their relationships on the glacier.

➤ **Zone of Accumulation:** In this region of the glacier, new snow is added. This location will be in the higher elevations, and when new material accumulates, it will slowly make its way down the slope due to gravity. We can think of this as the recharge area.

➤ **Zone of Ablation:** In this region of the glacier, the glacial ice melts and retreats. As the material added in the Zone of Accumulation makes its way down the elevation, it will melt in the Zone of Ablation. All glaciers have an ablation zone. Even in periods of extreme glaciation, there will always be a region where the glacier will be in a retreating phase.

➤ **Equilibrium Line:** As the name suggests, this is the line where the equilibrium is established between the accumulation and ablation zones. This is where melting and recharge is essentially balanced. During periods of healthy glaciation, the equilibrium line will extend to the lower elevations. The higher the elevation for the equilibrium line, the faster the glacier shrinks.

Glacial Erosion

The volume of ice and the speed with which the glacier moves toward sea level will ultimately determine the amount of erosion. Glaciers are a powerful force of erosion. They are slightly more effective than water, and can move larger-sized materials. One of the main

Figure 10.6 Image of glacial polish in artic Norway. All the landform features in this image were carved by glacial activity. The rocks in the forefront are smooth due to the abrasive activity of glaciation.

effects of glacial erosion is the polishing of the surface of the ground by abrasion. As glaciers move along the surface, they pick up material. This material will then act like sandpaper to wood as the glacier abrades the material along the bedrock below. This process of abrasion generates glacial polish and striations. In Figure 10.6, you can see some of the more rounded hills in the forefront. These gently sloping rock outcrops are the result of glacial polish. The smooth surfaces are worn down by the abrasive action of the glacier. In addition to glacial polish, this process will also generate striations. You should remember discussing striations back in the plate tectonic chapter. We learned about how Wegner used the directions of striations to argue for glacial movement. These striations are linear features that will be parllel with the flow of the glacier. Striations can be produced by alpine as well as continental glaciers. The transportation of material generates these striations (Figure 10.7). Glaciers can move materials the size of houses down to materials the size of flour. The process is slow, yet effective. In addition to transportation by ice, do not forget the meltwaters associated with glaciers. This can carry materials as well.

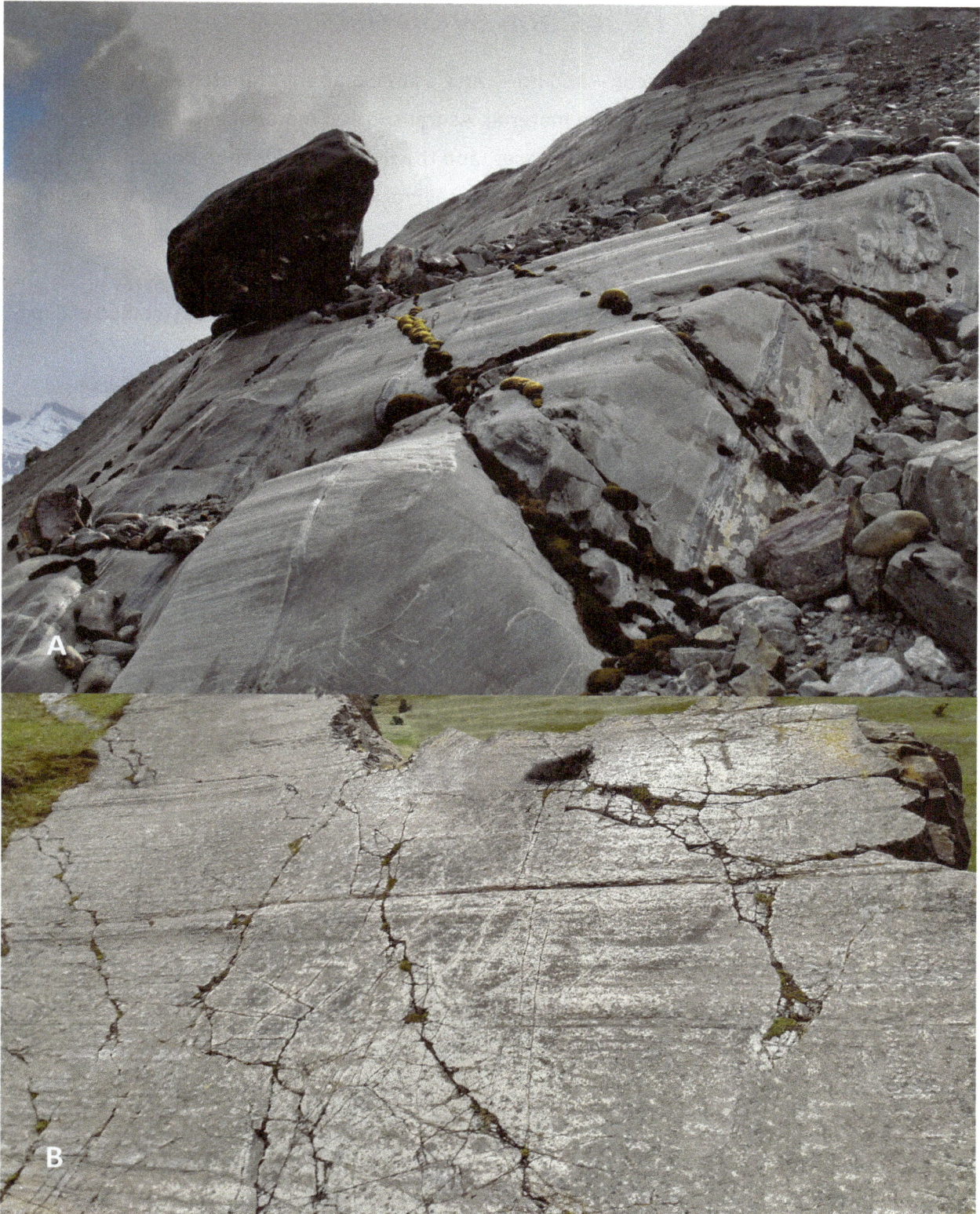

Figure 10.7 Examples of striations. A. Rock outcrop (South Georgia Island) has deep groves running parallel to the valley wall. B. Exposed surface (Edinburgh, Scotland) has a highly polished surface with slight striations. Both are the result of glaciation.

Deposition by Glaciers

Since we now know that glaciers can—and do—carry a large amount of materials, we must talk about what happens to this transported material. As transported material enters the zone of ablation, the rate of melting glacial ice increases, and the transported is then deposited. When it comes to material deposited by glaciers, we have several key terms that we need to keep in mind.

➤ **Drift:** Any material deposited by glaciers *or their meltwaters*. This implies that the material can be poorly sorted, and vary right up to the well-sorted. The key is in the method of deposition. Ice is a poor sorter, while meltwaters are an excellent sorter.

➤ **Till:** Material that is deposited directly by ice. This material is poorly sorted and arranged in the manner in which it was dropped. These two terms are similar (Figure 10.8). The best way to remember is to think that "All till is drift, but not all drift is till."

Figure 10.8   In this image, you can see the darker material on top of the glacier. If the ice melts and drops this material, it will be till and drift. If meltwaters carry the material off the glacier and deposit the material, it will be drift. Glaciers will carry material on, below, and within the ice.

➤ **Erratics:** Glacial erratics are enormous boulders transported and later deposited by melting glaciers. These erratics are often deposited far from their original source region. Erratics can be deposited by both alpine and continental glaciers (Figure 10.9).

Figure 10.9 The large boulder in this image is a glacial erratic. It was most likely transported down from the mountains above and dropped as the alpine glacier began to melt. Erratics will generally seem out of place in the surrounding landscape.

➤ **Moraines:** Linear features that are deposited at the bottom/end, along the sides, or found within glaciers. We will discuss moraines in more detail in the following pages.

Glacial Geomorphology

In this section, we are going to spend most of our time looking at alpine glacial geomorphology. The reason for this is that most of us will encounter this type of geomorphology more easily than any produced by continental glaciation. Since you are more likely to see these features, I want to make sure you recognize them. We are going to focus on four main geomorphological features.

➤ **Cirque:** A cirque (French) is an amphitheater-like or bowl-shaped valley formed by glacial erosion. An alternative name for this feature is **corrie** (Scottish Gaelic). In Figure 10.10, you can see a cirque/corrie. It is the hollowed out, bowl-like feature in the center of the alpine region. The second image is an example of what the cirque would have looked like before the glacier melted. As the glacier gathered in the bowl and moved to lower elevations due to gravity, it would have removed more and more material. This would make the bowl grow deeper and deeper over time. That is why many cirques also have lakes in them.

Figure 10.10 Image of a well-defined cirque. You can clearly see the bowl-like depression in the center of the higher elevations. This is located in high-latitude Norway. The lower green slopes below the cirque are most likely moraines. The cirque and the moraine were generated by an alpine glacier sitting in the bowl depression. Inset image is an artistic rendering of an alpine glacier eroding the cirque.

➤ **Arête:** a narrow ridge of rock that separates two valleys. It is typically formed when two glacial cirques erode toward one another.

➤ **Horn:** is an angular, sharply pointed mountain peak that results from cirque erosion due to multiple glaciers diverging from a central point. In Figure 10.11, you can see several horns that have been scoured by cirques.

Figure 10.11 Image of the Ritter Range in Northern California. You can see multiple horns that have been produced by cirques diverging from a centralized point. Image taken in June 2009.

➤ **U-Shaped Valley:** As glaciers scour down through previous V-shaped river valleys, they erode material as they go. This erosion changes the traditional V shape formed by river erosion to a gentler U-shaped valley generated by the broader bottom of the glacier. In the top image, to the right, you can see a modern-day, U-shaped valley. It does currently have a river, so it is transitioning back to a V-shaped valley. The below image shows what this same location would have looked like during the last glacial period, with a large alpine glacier carving its way down the valley to the sea. This U-shaped valley is located on the western coast of Scotland.

Moraines

Moraines are a common and important feature associated with glaciers. So, think about the glacier moving down its U-shaped valley. As the glacier moves along the valley walls, it picks up material. This material accumulates, and now, we have a band of sediments running along both sides of the glacier. This is what we call a **Lateral Moraine**. Keeping this in mind, now imagine what it might look like if this glacier joins another glacier as they both move down the valley. These two lateral moraines would join and they would no longer be on the edges, but would, rather, be in the middle. These are called **Medial Moraines**. In addition to picking up rocks and materials along the sides, the end of the glacier acts like a large bulldozer and scoops up material as it moves downhill. As long as it makes forward progress, this pile of material will grow. Once the glacier begins to retreat, it will leave this mound of material behind. We call this collection of material an **End Moraine**. The following image will show you where each type of moraine could be located.

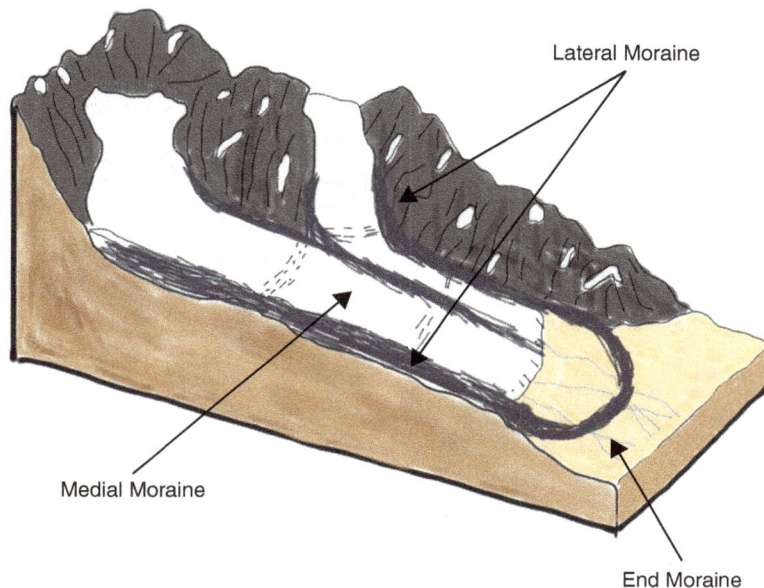

Lateral Moraine

Medial Moraine

End Moraine

Continental Glacier Geomorphology

Unfortunately, in the course of a lecture, we usually do not have the required time to spend on this topic. I would like to make it known that just because our class is skipping portions on continental glacier geomorphology, it does not mean that it is not a valuable and important aspect of glacial geology. You could take an entire course on continental glaciers and their resulting geomorphology. We spend our time on alpine geomorphology because that it is primarily what most of you will encounter. I do hope a basic understanding of glaciers will help you make informed decisions in the future.

Questions

1. What are the three primary regions of a glacier, and what happens in these zones?

2. How can we assess the health of an alpine glacier?

3. How are glaciers formed?

4. Where can you find continental glaciers?

5. What is the difference between Till and Drift?

6. What is a glacial erratic?

7. Describe a cirque.

8. What are the three types of moraines we discussed, and describe where each is located?

9. How are U-shaped valleys formed?

10. What are striations?

www.ingramcontent.com/pod-product-compliance
Lightning Source LLC
Chambersburg PA
CBHW081539220326
41598CB00036B/6485